SpringerBriefs in History of Science and Technology

More information about this series at http://www.springer.com/series/10085

Christopher D. Hollings

Scientific Communication Across the Iron Curtain

 Springer

Christopher D. Hollings
Mathematical Institute
University of Oxford
Oxford
UK

ISSN 2211-4564   ISSN 2211-4572   (electronic)
SpringerBriefs in History of Science and Technology
ISBN 978-3-319-25344-2   ISBN 978-3-319-25346-6   (eBook)
DOI 10.1007/978-3-319-25346-6

Library of Congress Control Number: 2015954967

Springer Cham Heidelberg New York Dordrecht London

Printed on acid-free paper

Springer International Publishing AG Switzerland is part of Springer Science+Business Media
(www.springer.com)

Dedicated to the memory of Jackie Stedall

Abstract

Drawing evidence from a range of disciplines, I study the extent to which scientists were able to communicate with their counterparts on the opposite side of what became the Iron Curtain. I consider the scope that existed for personal communication between scientists, as well as for the attendance of foreign conferences, and describe how these changed over the decades. Access to publications is also dealt with: I address separately the questions of *physical* access, and of *linguistic* access. In particular, I argue that physical accessibility was generally good in both directions, but that Western scientists were afflicted by greater linguistic difficulties than their Soviet counterparts, whose major problems in accessing Western research were bureaucratic in nature.

Keywords Scientific communication · Cold War science · Academic exchange

Preface

The various bureaucratic, logistical, and linguistic problems that afflicted contacts across the Iron Curtain are well-recognised by historians, and have been subjected to a number of general studies.[1] In the specifically scientific context, however, the vast majority of available investigations appear to focus upon those aspects of a particular discipline that emerged during the years of the Cold War, with communications difficulties often being a secondary matter. The purpose of the present book therefore is to provide an overview of the problems of Cold War scientific communications (principally in the academic context) in which these selfsame difficulties are the main focus. The account given here integrates (for the first time, to the best of my knowledge) the political/ideological/bureaucratic problems that afflicted East-West contacts with those of a linguistic nature. Indeed, my main theses lie at the intersection of these issues. They are, first of all, that, where conference attendance and personal correspondence were concerned, the difficulties encountered were broadly similar across the sciences, but that we begin to see a marked difference between distinct disciplines when we consider the availability of publications and the provision of translations. Second, I contend that the physical accessibility of publications from 'the other side' was generally much better than is commonly supposed. Finally, I argue that Western scientists were afflicted by greater linguistic difficulties than their Soviet counterparts, who do not appear to have been overly affected by the language barrier—instead, their major problems in accessing Western research were bureaucratic in nature.

The material of this book is arranged as follows. I begin in Chap.1 by establishing a rough framework for the study of scientific communication, and also set up the conventions and terminology to be adopted throughout. In Chap. 2, I turn to a general discussion of communications between scientists in East and West from the 1920s, up to around the 1980s; the focus in Chap. 2 is upon *personal* contacts between scientists, by which I mean correspondence and face-to-face meetings, usually at conferences. The scene is set in Sect. 2.1 by a short discussion of

---

[1] I save detailed references for the Introduction.

East-West scientific communications in the years before the First World War. Sections 2.2 and 2.3 then take the 1920s and 1930s in turn. We will see that contacts were initially quite easy, though they diminished under Stalin. Section 2.4 examines the slightly improved communications that took place during the Second World War, in the spirit of wartime cooperation, but which were then stifled in the early Cold War climate (Sect. 2.5). Following Stalin's death in 1953, contacts were opened up once again, as I discuss in Sect. 2.6, but these were not without their problems. A comprehensive account of the political difficulties afflicting Soviet scientists during this period (with particular regard to international contacts) is provided by the writings of the biologist Zh. A. Medvedev, which we examine in Sect. 2.7. Although such problems were largely a concern for *Soviet* scientists, there were nevertheless some home-grown political problems that affected Western scientists—I discuss these in Sect. 2.8. The material on personal communications is brought to a close with some concluding remarks in Sect. 2.9.

Issues surrounding physical access to publications are addressed in Chap. 3. After some general remarks in Sect. 3.1, I consider the matter of censorship (Sect. 3.2), and the tendency of Soviet scientists to publish much of their work in 'local' journals, which often did not find their way into Western libraries (Sect. 3.3). Efforts to gain broad impressions of the work of 'the other side' are dealt with next: abstracting services are the subject of Sect. 3.4, whilst Sect. 3.5 looks at the many published Western surveys of Soviet scientific advances.

In Chap. 4, we turn to linguistic matters, which I set in the context of the infamous 'foreign-language barrier' (Sect. 4.1). The specific issues considered here are the use of foreign languages (Sect. 4.2), and the appearance of foreign authors (Sect. 4.3), in Soviet journals, Russian-language ability amongst Western scientists (Sect. 4.4), and the translation of scientific works (Sect. 4.5).

Some final remarks, and points to be pursued, may be found in Chap. 5. One of my goals with this book was to achieve something approaching comprehensiveness in the resources cited throughout. It is therefore my hope that these references, in conjunction with the comments in the Conclusion, will provide the impetus for further research in this area.

The present text emerged as a much-expanded version of Chap. 2 of my book *Mathematics across the Iron Curtain: A History of the Algebraic Theory of Semigroups* (American Mathematical Society, Providence, RI, 2014). However, whereas that chapter focused on communications in *mathematics* (indeed, in many places, it did not stray very far from abstract algebra), the present text adopts a much broader point of view, taking in as many different disciplines as possible. Nevertheless, the reader might still detect a slight bias towards mathematics. Given its origins, this book contains, at its core, some research carried out between March 2010 and February 2013 at the Mathematical Institute of the University of Oxford under the auspices of research project grant number F/08 772/F from the Leverhulme Trust. Thanks must finally go to the two anonymous referees whose insightful comments have not only helped to improve this book, but have also suggested directions for future research.

Contents

Chapter 1
Introduction and Overview

Abstract In this chapter, we set the scene for our subsequent study by briefly considering previous work on Cold War scientific communication. In addition, we outline the main themes to be found in later chapters.

Keywords Scientific communication · Cold War science · Academic exchange

In order to set this study in motion, I first pick up on the comments made in the Preface concerning previous studies of Cold War (scientific) communication. There have been several such investigations, ranging in focus from East-West exchanges of a broadly-defined cultural nature (Richmond 2003; David-Fox 2012) to those of a more overtly scientific type (Krementsov 2005; Schweitzer 1989, 1992, 2004), taking in general academic exchanges along the way (Byrnes 1962, 1976). Such studies stand alongside research on the broad topic of 'Cold War science',[1] and also, more specifically, of Soviet science.[2] As noted in the Preface, some details regarding Cold War scientific communication may also be found in various historical investigations of specific disciplines. To take the study of 'Cold War mathematics' as an example, the literature contains a range of works that touch, if only briefly, upon the problems of international communications during the relevant period.[3] In the present text, I hope to provide a concise general study that will complement, and perhaps even begin to unite, the various resources cited above.

I adopt a very broad perspective, taking in as many branches of science as possible; in the process, I hope to offer a new perspective on Cold War scientific communications, and, moreover, to provide a framework within which further research in this area might be conducted. My approach is very 'examples-led': in dealing, for instance, with the presence (or otherwise) of Soviet scientists at conferences held

---

[1] See, for example, the articles in vol. 31 (2001) of *Social Studies in Science*, vol. 101 (2010) of *Isis*, and vol. 55, no. 3 (2013) of *Centaurus*; see also Krementsov (2002) and Wolfe (2013).

[2] See, for example, Birstein (2001), Graham (1972, 1993, 1998), Krementsov (1997), Lubrano and Solomon (1980), and Pollock (2006). See also the resources cited in note 28 on p. 66.

[3] See, for example, Gerovitch (2002), Gessen (2011), Graham and Kantor (2009), Hollings (2014), and Zdravkovska and Duren (1993). Siegmund-Schultze (2014), on the other hand, *does* deal directly with the issue of international mathematical communication.

© The Author(s) 2016
C.D. Hollings, *Scientific Communication Across the Iron Curtain*,
SpringerBriefs in History of Science and Technology,
DOI 10.1007/978-3-319-25346-6_1

in the West, I have endeavoured to draw upon the reports of a wide range of such meetings. Amongst my main sources for such reports have been the journals *Science* and *Nature*, which are extremely useful also for contemporary views of the various issues discussed here. Although I have used some Soviet sources, the vast majority of the materials that I cite are Western in origin—this is partly a matter of accessibility (perhaps somewhat ironically, given the content of Chap. 3), and partly the fact that Western sources on this subject appear to be more numerous, and tend to have rather more to say.

Before we proceed further, certain elements of terminology need to be addressed. First of all, my general use of the term 'Iron Curtain', particularly in the title, is somewhat anachronistic: the Iron Curtain, as the term is generally understood, did not come into existence until 1945, and yet my main period of study begins rather earlier, in around 1917. I hope that the reader will excuse my slightly inaccurate use of 'Iron Curtain' to mean simply a divide between communist Central and Eastern Europe and the West. With regard to the terms 'East' and 'West', the latter will usually refer to Western Europe and North America (with a strong bias towards English-speaking scientists), whilst the former will refer here only to the USSR. I hope eventually to deal with the other communist-bloc countries elsewhere. In addition, I ought to note that in Chap. 2 (and also, to some extent, in Chap. 3), I will have rather more to say about the situation in the Soviet Union than about that in the West. This is because I believe (and will argue) that there is in fact more to say: that the greater part of the politically- or bureaucratically-motivated bars to communication originated in the USSR. In Chap. 4, the focus will be reversed.

Levels of contact between scientists on opposite sides of the Iron Curtain varied over the decades. Early in our period of interest (from about 1917 to the mid-1920s), communications problems were caused successively by the First World War (1914–1918), the October Revolution (1917), and the Russian Civil War (1917–1922), before levels of contact returned to something like those prior to 1914: scientists travelled freely in and out of the newly-formed Soviet Union, and, as the 1920s progressed, access to scientific publications was similarly enhanced. These steady improvements were halted, however, by Stalin's rise to power in the late 1920s. Around this time, increased demands for 'ideological orthodoxy' translated into criticism of those many Soviet scientists who chose to publish their work abroad, and state suspicion of anyone with unsanctioned foreign contacts. The result was the wholesale isolation of Soviet science, at least as far as personal communications and conference attendance were concerned—printed works still seem to have got through, to an extent that I will discuss in Chap. 3.

Over the course of the 1930s, the levels of communication between scientists in East and West dwindled, though not quite to zero. Nevertheless, the situation remained difficult until Stalin's death in 1953. During the subsequent 'thaw' under Khrushchev, international (scientific) contacts improved once more, with Western scientists availing themselves of new-found opportunities to visit the USSR. Travel in the opposite direction was also possible, although it was very often hindered by bureaucratic obstacles, as we will see. In addition, the Soviet authorities remained suspicious of those citizens who had, or attempted to have, any contact with foreigners. Similar suspicions existed within the West (particularly in the United States), though on a considerably smaller scale. Thus, at the start of the 1960s, international scientific exchanges were increasingly becoming possible, but they were by no means free of obstacles.

Although the ability of scientists to travel across the Iron Curtain varied over time, the provision of published materials was, by and large, good throughout the relevant decades, as I have already asserted and will argue in more detail in due course. Indeed, the exchange of printed matter was often the only point of contact between scientists in East and West. Although coverage was certainly not comprehensive, a broad selection of Western scientific publications appears to have been available in the academic libraries of the USSR, and vice versa. A lack of knowledge of scientific work on the other side of the Iron Curtain thus cannot simply be blamed on the inaccessibility of publications (although this was sometimes the case). In order to approach a fuller understanding of this matter, we must turn to the issue of language.

As one studies Soviet science, one gains the impression that, in general, the scientists of the USSR had an understanding of at least one or two of 20th-century Western science's three principal languages: English, French and German. Indeed, it is very rare to encounter any suggestion of language difficulties when considering Soviet scientists' efforts to learn of Western research. The situation in the opposite direction was of course rather different: a typical Western scientist was/is likely to have at least a reading knowledge of English, French or German, but was/is probably not able to handle Russian quite so well—something of a problem when, as we will see, Soviet scientists were under pressure to confine their output to Russian-language journals. The physical accessibility of a Western source in a Soviet library would probably have been sufficient to make it accessible to the Soviet reader (issues of possible censorship aside), but the same cannot be said of the opposite situation. We can see this by considering Western scientific commentaries and personal reminiscences from the middle decades of the 20th century—these contain numerous comments regarding language problems: see, for example, those quoted in Hollings (2014, pp. 12,38).

The learning of sufficient Russian to be able to keep abreast of Soviet scientific advances seems to have been too much for some Western scientists, whilst others questioned the need: was there anything in the Soviet literature worth seeing? Opinions in this connection varied from discipline to discipline, as we will see later on, but in many instances,

[the] appreciation of Russian science and technology by non-Russians was obstructed by linguistic barriers, ethnic prejudices, and simple ignorance. (Graham 1972, p. 16)

Such barriers were generally to the detriment of science (and, indeed, to the purses of funding bodies), since the almost inevitable result of ignorance of the work of 'the other side' was the duplication of research. One example of this is provided by the parallel work of Norbert Wiener (USA) and A.N. Kolmogorov (USSR) in cybernetics (Gerovitch 2002, p. 58), whilst another, moving beyond the Soviet context, lies in the independent work carried on the structure of haemoglobin by biologists in Czechoslovakia and the USA (Medvedev 1971, p. 116). Within mathematics, this duplication often manifests itself in the form of double-named theorems.[4] Although the occasional allegation of plagiarism has surfaced (in the form of both vague innuendo and direct accusation[5]), such duplications have usually been recognised as a lamentable side-effect of communications problems: see, for example, the comments in Hollings (2014, p. 13).

One final issue to be addressed here is that of scientific internationalism. It is often observed that scientists enjoy a certain 'cultural contiguity', which transcends national borders, and leads to a desire for open international collaboration, regardless of the relative political positions of national governments. However, it would be rather naïve to blame any breakdowns in international scientific communication simply on the actions of governments, for scientists have certainly been political players themselves over the decades.[6] Indeed, the idea that an internationalist stance is science's natural state has been challenged by some authors,[7] who argue, moreover, that the study of the history of science has been blighted by the "universalist presupposition" (Crawford 1992, p. 1). Having acknowledged this issue, however, I set it aside. Although I do not argue that internationalism is natural to science, I nevertheless use it as a convenient benchmark against which to measure the levels of communication that are to be discussed here.

References

Anon: Ohio inquiry ordered on plagiarism charge. NY Times, 31 Dec (1986)

Anon: Follow-up on the news; Plagiarism charge against professor. NY Times, 26 Apr (1987)

Birstein, V.J.: The Perversion of Knowledge: The True Story of Soviet Science. Westview Press, Boulder, CO (2001)

Byrnes, R.F.: Academic exchange with the Soviet Union. Russian Rev. 21(3), 213–225 (1962)

Byrnes, R.F.: Soviet–American Academic Exchanges, 1958–1975. Indiana Univ. Press (1976)

Crawford, E.: Nationalism and Internationalism in Science, 1880–1939: Four Studies of the Nobel Population. Cambridge Univ. Press (1992)

David-Fox, M.: Showcasing the Great Experiment: Cultural Diplomacy and Western Visitors to the Soviet Union, 1921–1941. Oxford Univ. Press (2012)

Gerovitch, S.: From Newpeak to Cyberspeak: A History of Soviet Cybernetics. MIT Press (2002)

Gessen, M.: Perfect Rigour: A Genius and the Mathematical Breakthrough of the Century. Icon Books (2011)

Graham, L.R.: Science and Philosophy in the Soviet Union. Alfred A, Knopf, New York (1972)

Graham, L.R.: Science in Russia and the Soviet Union: A Short History. Cambridge Univ. Press (1993)

Graham, L.R.: What Have We Learned About Science and Technology from the Russian Experience? Stanford Univ. Press (1998)

---

[4]See, for example, the comments of Gessen (2011, p. 7).

[5]For instances of the former, see Lohwater (1957) or Rathmann (1958); for the latter, see Rich (1986, 1987), Shabad (1986), and Anon (1986, 1987).

[6]In the US context, see, for example, Kuznick (1987).

[7]See, for example, Crawford (1992, Chap. 2) or Somsen (2008).

Graham, L., Kantor, J.-M.: Naming Infinity: A True Story of Religious Mysticism and Mathematical Creativity. The Belknap Press of Harvard Univ. Press (2009)

Hollings, C.: Mathematics across the Iron Curtain: A History of the Algebraic Theory of Semigroups. Amer. Math. Soc., Providence, Rhode Island (2014)

Krementsov, N.: Stalinist Science. Princeton Univ. Press (1997)

Krementsov, N.: The Cure: A Story of Cancer and Politics from the Annals of the Cold War. Univ. Chicago Press (2002)

Krementsov, N.: International Science between the World Wars: The Case of Genetics. Routledge, New York and London (2005)

Kuznick, P.J.: Beyond the Laboratory: Scientists as Political Activists in 1930s America. Univ. Chicago Press (1987)

Lohwater, A.J.: Mathematics in the Soviet Union. Science 125(3255), 17 May, 974–978 (1957)

Lubrano, L.L., Solomon, S.G. (eds.): The Social Context of Soviet Science. Westview Press, Boulder CO (1980)

Medvedev, Zh.A.: The Medvedev Papers: The Plight of Soviet Science Today. Macmillan (1971)

Pollock, E.: Stalin and the Soviet Science Wars. Princeton Univ. Press (2006)

Rathmann, F.H.: Soviet scientific literature. Science 128(3325), 19 Sept, 678 (1958)

Rich, V.: Plagiarism charges levelled. Nature 324(6094), 20 Nov, 198 (1986)

Rich, V.: Row takes new turn over US plagiarism of Soviet books. Nature 25(6107), 26 Feb, 748 (1987)

Richmond, Y.: Cultural Exchange and the Cold War: Raising the Iron Curtain. Pennsylvia State Univ. Press (2003)

Schweitzer, G.E.: Techno-diplomacy: US-Soviet Confrontations in Science and Technology. Plenum Press, New York and London (1989)

Schweitzer, G.E.: US-Soviet Scientific Cooperation: The Interacademy Program. Technol. in Soc. 14, 173–185 (1992)

Schweitzer, G.E.: Scientists, Engineers, and Track-Two Diplomacy: A Half-Century of U.S.-Russian Interacademy Cooperation. Nat. Acad. Press, Washington DC (2004)

Shabad, T.: Soviet scholars say American plagiarized; he defends himself. NY Times, 29 Dec (1986)

Siegmund-Schultze, R.: One hundred years after the Great War (1914–2014): a century of breakdowns, resumptions and fundamental changes in international mathematical communication. In: Jang, S.Y., Kim, Y.R., Lee, D.-W., Yie, I. (eds.) Proceedings of the International Congress of Mathematicians Seoul 2014, vol. IV, pp. 1231–1253. Gyeong Munsa (2014)

Somsen, G.J.: A history of universalism: conceptions of the internationality of science from the Enlightenment to the Cold War. Minerva 46, 361–379 (2008)

Wolfe, A.J.: Competing with the Soviets: Science, Technology, and the State in Cold War America. Johns Hopkins Univ. Press (2013)

Zdravkovska, S., Duren, P.L. (eds.): Golden Years of Moscow Mathematics. Amer. Math. Soc./London Math. Soc. (1993); 2nd ed. (2007)

Chapter 2
Personal Communications

Abstract In this chapter, we provide a general discussion of communications between scientists in East and West from the 1920s, up to around the 1980s, with the focus being upon *personal* contacts between scientists: correspondence and face-to-face meetings. We will see that the initially quite easy contacts of the 1920s became rather more difficult under Stalin, before picking up again slightly during the Second World War, and then more dramatically following Stalin's death.

Keywords Scientific communication · Academic exchange programmes · Wartime exchange · Zhores Medvedev · McCarthyism · International congresses

2.1 Before the First World War

Before embarking upon a study of the extent to which Russian scientists were able to communicate with the rest of the world (the West in particular) during the Soviet era, it will first be beneficial to gain some idea of the situation in earlier decades, namely, in those immediately prior to the First World War.

The closing years of the 19th century and the early ones of the 20th were a period of great growth in the international practice of science: these were the years in which several international scientific organisations were founded, and also when the first discipline-specific international congresses were held.[1] According to some estimates, there were around 20 international scientific congresses per year during the last three decades of the 19th century, and around 30 per year in the 15 years prior to the First World War (Crawford 1992, p. 55). Moreover, the number of international scientific organisations is said to have doubled every ten years from 1885 onwards (though, in the long run, with a mortality of 60 %: see Crawford 1992, pp. 40–41).

---

[1] For a general overview of this growth in international scientific activities, see Rosenzweig (2000, Chap. 2); see also Crawford (1992, pp. 38–43).

© The Author(s) 2016   7
C.D. Hollings, *Scientific Communication Across the Iron Curtain*,
SpringerBriefs in History of Science and Technology,
DOI 10.1007/978-3-319-25346-6_2

The Russian Empire of this period played its part in these activities, and certainly had scientists of world renown[2]: we might mention P.L. Chebyshev (1821–1894), D.I. Mendeleev (1834–1907), N.E. Zhukovskii (1847–1921), I.P. Pavlov (1849–1936), and V.I. Vernadskii (1863–1945). Indeed, the fact that several of these men received honours from foreign scientific societies (for example, Mendeleev's receipt of the Royal Society's Copley Medal, and Pavlov's of the Nobel Prize in Physiology or Medicine[3]) indicates further that Russian science was taken seriously on the world stage—a consideration that will become relevant as we move into the Soviet era.

A perusal of scientific sources from this period suggests that communications between Russian scientists and those of the rest of the world were, to put it loosely, as good as we might expect them to have been, given the technologies and postal services of the time. The Russian Imperial Academy of Sciences was, for example, a member of the International Association of Academies (IAA), which existed between 1889 and 1914 (see Greenaway 1996, Chap. 1). Indeed, the Association's 1913 General Assembly was held in Saint Petersburg (Greenaway 1996, p. 14). Travel to and from Russia appears to have been hindered only by expense and by the limitations of the transport provisions of the era. To take the Russian scientists on the above list as cases in point, we note, for example, that Mendeleev worked for some time in Heidelberg (Gordin 2008; Graham 1993, p. 48), and Pavlov in Leipzig (Graham 1993, p. 239); Vernadskii studied both in Naples and in Munich (Kautzleben and Müller 2014). Indeed, the relevant literature is full of many other instances of Russian scientists travelling abroad in the late-19th and early-20th centuries.

The ability to travel in order to work or to study, however, is a slightly different matter from the provision of the type of travel involved in scientific communication. In order to gauge the latter, we need to consider Russian attendance at international conferences. Let us take, for example, the International Congresses of Mathematicians (ICMs), the first of which was held in Zurich in 1897.[4] As we can see from Table 2.1, the number of Russian and Ukrainian delegates at this first congress was small, but, out of the 16 countries from which the congress' various attendees hailed, the Russian presence was ranked 6th in terms of its size, after Switzerland, Germany, France, Italy and Austria-Hungary (Rudio 1897, p. 78). Indeed, these five countries, together with Russia, were the only nations whose contingents numbered in double figures. Looking further down Table 2.1, we see that Russian mathematicians

---

[2]For succinct overviews of Russian science in this period, see Krementsov (1997, 2006); for a comprehensive account, see Vucinich (1970).

[3]On Mendeleev, see Hargittai et al. (2007); on Pavlov, see Graham (1993, p. 239). For comments on Mendeleev's foreign contacts, and for a discussion of the status of 19th-century Russian chemistry, see Gordin (2015, Chaps. 2 and 3).

[4]Levels of Russian/Soviet attendance at the ICMs, as recorded in Table 2.1, will be used for illustrative purposes throughout this chapter. I have compiled these figures by using the various congress proceedings and other sources, and some of the numbers given differ from those that appear in previous books and articles on the ICMs. In the interests of saving space, I do not explain these figures here, but I hope to do so elsewhere.

Table 2.1 Soviet attendance at International Congresses of Mathematicians (ICMs), 1897–1990 (prior to 1917, the number of Russian and Ukrainian delegates is given); figures compiled from contemporary reports, congress proceedings, and Lehto (1998)

Year	Location	Total number of delegates	Number of 'Soviet' delegates
1897	Zurich	208	13
1900	Paris	253	11
1904	Heidelberg	336	22
1908	Rome	535	16
1912	Cambridge	574	23
1920	Strasbourg	200	1
1924	Toronto	444	5
1928	Bologna	836	37
1932	Zurich	667	3
1936	Oslo	476	0
1950	Harvard	1,700	0
1954	Amsterdam	1,553	5
1958	Edinburgh	1,658	32
1962	Stockholm	2,107	42
1966	Moscow	4,277	1,479
1970	Nice	2,810	129
1974	Vancouver	3,121	50
1978	Helsinki	3,042	55
1983	Warsaw	2,233	283
1986	Berkeley	3,586	57
1990	Kyoto	4,102	110

maintained a presence at these congresses until the First World War. Indeed, Russia continued to be one of the few countries that provided a number of delegates in double figures.[5]

With regard to other disciplines, we see, for example, that four Russians (Mendeleev amongst them) and one Ukrainian attended the International Congress of Chemists in Karlsruhe in 1860 (Wurtz 1929).[6] Moving a little closer to our period of interest, Russian chemists appear also to have had a small presence at the subsequent International Congresses of Applied Chemistry (Wiley 1896, p. 923). Indeed, had the First World War not intervened, the ninth such congress would have been held in Saint Petersburg in 1915 (Burns and Deelstra 2011, p. 281). To take some other late-19th-century international conferences,[7] Russia was strongly represented at the First International Congress of Physiologists in Basel in 1889 (Franklin 1938, pp. 246, 328), and also at the Sixth International Geographical Congress,

---

[5]On Russian attendance of the early ICMs, see Demidov and Tokareva (2005, pp. 144–145).

[6]On this congress, see also Milt (1951) and Ihde (1961).

[7]For a list of late-19th- and early-20th-century scientific conferences, see Baskerville (1910).

held in London in 1895 (Keltie and Mill 1896, Appendix A). Indeed, Russian delegates had attended the London Geological Congress seven years earlier (Anon 1888). The first two international psychological congresses of the 20th century (Paris, 1900; Rome, 1905) were attended by Russian delegates (Rosenzweig 2000, pp. 35, 37), as were the Fourth International Genetics Congress (Paris, 1911: see Krementsov 2005, p. 4) and the First International Eugenics Congress (London, 1912: see Krementsov 2005, pp. 16–17). Thus, it appears from these few examples that Russian scientists were playing a very active role in world science in this period, and, in spite of the distances that they often had to travel, were frequent attendants of foreign conferences.

I conclude this section with a few brief comments on the opposite consideration: foreign travellers in Russia, and foreign participation in conferences held there. As for Russian scientists abroad, the historical literature is full of examples of non-Russians travelling to Russia for scientific visits.[8] Formal international conferences, on the other hand, appear to have taken place in Russia far less frequently than they did in, say, Britain, France or Germany, but they nevertheless occurred.[9] I cite, for example, the International Congresses on Anthropology, Prehistoric Archaeology and Zoology, held in Moscow in 1892 (Anon 1893; Sommer 2009),[10] the International Geological Congress, involving various expeditions across Russia, in 1897 (Palache 1897; Milanovsky 2004), and also the Eleventh International Congress of Navigation (Saint Petersburg, 1908: see Congrès 1908, 1910), all three of which attracted large numbers of foreign delegates. We see then that, although the major Russian cities were rather remote from the other scientific centres of Europe (and certainly from those of North America), conferences held in Russia in the few decades before the First World War were nevertheless able to boast significant numbers of attendees from other countries.

2.2 The 1920s

The First World War naturally had an enormously disruptive influence on international communications, and on international scientific contacts in particular, though I do not attempt to go into this here (see instead Kevles 1971). In the case of Russia, the October Revolution of 1917 and the civil war of 1917–1922 effected further obstruction, and yet, during the 1920s, communications between scientists in the newly-formed USSR and those in the West began slowly to return to something like their pre-war levels. In Russia, the Academy of Sciences spearheaded efforts

---

[8]To take some arbitrary examples: the American physiologist Francis Gano Benedict (see Neswald 2011, 2013), and a number of Spanish physicists (Sánchez-Ron 2002).

[9]Baskerville (1910) lists over 150 international congresses in the sciences, humanities and arts, but only two that took place in Russia.

[10]Indeed, Russian delegates regularly attended the early International Archaeological Congresses; see Marton (2009).

to re-establish contacts with the scientists of other nations (Rich 1974; see also Strekopytov 1977). However, the initial reluctance of some Western powers to recognise the Soviet Union hampered scientific exchanges through government channels for some time (Furaev 1974). Moreover, the USSR was not invited to join the newly-formed International Research Council (the successor to the IAA (p. 8), and predecessor of the International Council of Scientific Unions, ICSU).[11] Nevertheless, these difficulties were offset somewhat by the efforts of individuals (particularly in the provision of printed matter—see Sect. 3.1). Indeed, this was the period of the 'fellow travellers': the (often uncritical) Western enthusiasts for the nascent Soviet Union who made their 'pilgrimage to Russia'. Estimates suggest that around 100,000 Americans and Europeans visited the USSR during the 1920s, scientists amongst them (David-Fox 2012, p. 1).

German-Soviet scientific ties appear to have been particularly prominent during this period (as, indeed, they had been before the First World War), since both countries found themselves largely excluded from international scientific activities.[12] Some Russian scientists of the 1920s were also in receipt of funding from the US-based Rockefeller Foundation[13]—the attitude of some Americans appears to have been that even in the absence of formal diplomatic ties with the fledgling USSR, US influence might nevertheless be increased through patronage (Hamblin 2000a).

During the early years of the USSR's existence, there do not appear to have been any particularly stringent restrictions on Soviet scientists with regard either to correspondence or to travel, although the requirement for both Ministry of Education and Central Committee approval for any foreign trips was instituted in 1924 (Solomon and Krementsov 2001, p. 275): permission to travel abroad became a source of power for the Communist Party over the intelligentsia (David-Fox 2002). However, as the Georgian-born biologist Zh.A. Medvedev (for more on whom, see Sect. 2.7) later commented in his book *Soviet Science*,

> [t]he shortage of foreign currency was the main factor limiting the opportunities for official foreign travel. (Medvedev 1979, p. 16)

Nevertheless, Soviet scientists were able to travel throughout Europe, and even attended conferences in North America. The Russian engineer and applied mathematician A.N. Krylov, for example, addressed the 1921 Edinburgh meeting of the British Association for the Advancement of Science (Anon 1921). The pure mathematician P.S. Aleksandrov, on the other hand, travelled widely in continental Western Europe during the 1920s (Aleksandrov 1979), as did the geneticist N.I. Vavilov (who also travelled to the USA: see Krementsov 2005, pp. 23–24), and a delegation of Russian physicists. Headed by A.F. Ioffe, this

---

[11] See Greenaway (1996, p. 18) or Schroeder-Gudehus (1973, p. 115).

[12] See Solomon and Krementsov (2001, pp. 276–277, 287), Forman (1973, pp. 167–168) and Schroeder-Gudehus (1973, p. 115). On the scientific ostracism of Germany (and the other Central Powers) more generally, see Cock (1983) or Crawford (1988).

[13] See Solomon and Krementsov (2001), Kojevnikov (1993) or Kojevnikov (2004, Sect. 4.2). For more on the Rockefeller Foundation's sponsorship of European scientists during this period (with a particular focus on mathematicians), see Siegmund-Schultze (2001).

last group set out in 1921 "with the goal of purchasing equipment and scientific literature" (Kojevnikov 2004, p. 102). Amongst Ioffe's travelling companions was the physicist P.L. Kapitsa, who subsequently worked at the Cavendish Laboratory in Cambridge from 1921 to 1934 (Kojevnikov 2004, Chap. 5). Foreign scientific visitors also found their way to Russian universities during this period.[14] From 1925 onwards, any such visitors were aided during their stay in the USSR by representatives of the All-Union Society for Cultural Relations with Foreign Countries (VOKS/ВОКС = Всесоюзное общество культурн ой связи с заграницей), whose mission, carefully crafted to give the appearance that it was not under direct state control (David-Fox 2002, pp. 11, 26), was to present the Soviet Union in an entirely positive light.[15] This is particularly relevant for our purposes, since, as Susan Gross Solomon and Nikolai Krementsov have observed,

> [s]cience was a major focus of VOKS's brief: the bulletin of VOKS, issued in French, German, and English, trumpeted Soviet scientific advances.[16]

VOKS established links with 'friendship societies' overseas (David-Fox 2012, Chap. 2), including, for example, a French-Soviet rapprochement society (Stern 1997, 1999),[17] the National Council of American-Soviet Friendship,[18] and the Society for Cultural Relations between the Peoples of the British Commonwealth and the USSR (SCR),[19] the latter two of which had dedicated science sections.[20] Indeed, in the case of the SCR, the science section seems to have been one of the most active parts of the society: the initiative for certain specialised trips to the USSR came from the science section (Lygo 2013, pp. 587–588).

With regard to conferences, we can look again to Table 2.1. After the attendance of the 1920 ICM by just one Russian (in fact, an exile, D.P. Riabouchinsky, then resident in the south of France: see Villat 1921, p. xiii), Soviet involvement in the congresses began to increase. Russian scientists were also represented at foreign conferences in other disciplines: take, for example, their presence at each end of the decade at both the Tenth (Paris, 1920) and Thirteenth (Boston, 1929) International Congresses of Physiologists (Franklin 1938, pp. 293, 305), and also the First International Congress of Soil Science, held in Washington, DC, in 1927 (Deemer et al. 1928). Indeed, Soviet attendees formed the largest foreign delegation at the Fifth International Genetics Congress in Berlin in 1927 (Krementsov 2005, pp. 4, 19). In contrast, however, no Soviets were present at the Seventh International Congress of Psychology in Oxford in 1923 (Rosenzweig 2000, p. 44). Nevertheless, as the

---

[14]The references given in note 8 on p. 10 are again relevant here.

[15]See Solomon and Krementsov (2001, p. 287), David-Fox (2012, Chap. 1) or, for an older source, Kameneva (1928).

[16]Solomon and Krementsov (2001, p. 287); on the bulletin, see also David-Fox (2012, p. 90).

[17]On Soviet-French ties, see also Ivanovskaya (1976), Plaud (1980) and Fedorov (1984).

[18]For a quite critical history of this society, see Nemzer (1949, pp. 275–279).

[19]See Lygo (2013), and also King (1967) and Todd (1967).

[20]See Lear (1997, p. 262) or Lygo (2013, p. 584); on the American-Soviet Science Society (established in 1944), see Krementsov (1996, p. 240).

1920s progressed, more and more Soviet scientists were sent abroad at the expense not only of the Academy of Sciences (Medvedev 1979, p. 16), but also of a number of government ministries, such as the Ministry of Education (with jurisdiction over 'pure' science) and the Supreme Council of the National Economy ('applied' science) (Solomon and Krementsov 2001, pp. 275, 285–286). In their efforts to boost foreign contacts, these ministries founded special departments devoted to foreign relations, and also established networks of foreign representatives (Solomon and Krementsov 2001, p. 276).

In 1925, the Academy of Sciences staged an international conference to commemorate its 200th anniversary—the first international scientific meeting to be held in Russia since the start of the First World War, and a conscious attempt to re-establish scientific ties (Sorokina 2006, pp. 63, 65). The conference was attended (at Soviet expense) by many foreign scientists,[21] from a range of disciplines; definitive attendance figures seem to be lacking, but the number of foreign attendees could have been as high as 98 (Bateson 1925). A report of the meeting subsequently appeared in *Nature*, penned by one of the British delegates, the geneticist W. Bateson. He considered that the conference "had been organised largely with an eye to its propaganda-value" (Bateson 1925, p. 681), but his report suggests that the foreign delegates enjoyed free interaction with their Soviet colleagues. A reciprocal conference was held in London later the same year (Anon 1925). Both meetings received a very enthusiastic write-up in a book produced 50 years later (in English) by the Soviet Academy of Sciences: *USSR Academy of Sciences: Scientific Relations with Great Britain* (Korneyev 1977).[22] This volume records, for example, the many positive remarks that were supposedly made by British scientists concerning Soviet scientific organisation, and the USSR more generally—Western comments of a more critical type, such as those made by Bateson in his report ("Of liberty we saw no sign": Bateson 1925, p. 683), do not appear. Indeed, this book is quite typical of the Soviet-produced sources that deal with international scientific communications, in that it omits anything that might show up the USSR in a negative light—any communications difficulties are blamed, often in a somewhat hysterical tone, on the 'reactionary' attitudes of Western governments. For example, an article in this volume refers to the "pronounced anti-Soviet attitude" of the "ruling clique of Britain" (Korneyev and Timofeyev 1977, p. 9),[22] before going on to assert the interest in all things Soviet shown by the British people at large, and by British scientists in particular, who

> recognised the Soviet state much earlier than the government and who, being interested in scientific contacts as much as their Soviet colleagues, played a substantial role in breaking

---

[21] Including German delegates, at a time when German scientists were largely excluded from international meetings; see Forman (1973, p. 168). Indeed, the Germans formed the largest foreign contingent (Sorokina 2006, p. 64); for a list of countries who sent delegates, see Sorokina (2006, p. 85).

[22] Korneyev's initials are given as S.C. on the title page of Korneyev (1977), but as S.G. in the article Korneyev and Timofeyev (1977). One of these sets of initials is certainly wrong, but I do not know which.

through the anti-Soviet blockade and in surmounting the psychological barrier erected by
the campaign of slander which was being whipped up in Britain.[23]

As we will see later, there was a marked tendency amongst Soviet commentators to
overstate the levels of contact that were possible, particularly in later decades.

Thus, during the 1920s (particularly in the second half of the decade), scientists on both sides of what later became the Iron Curtain appear once again to have
enjoyed a level of communication comparable to that before the First World War.
I have given just a few documented examples of Soviet scientists travelling to the
West, and of travel in the opposite direction—a glance through the proceedings of
many international conferences of the 1920s would undoubtedly uncover further
instances.[24]

2.3 The 1930s

As far as the issue of international scientific communication was concerned, the
1930s had a promising start, with 14 foreign delegates present at the 1930 First
All-Union Congress of Mathematicians in Kharkov, six of whom delivered talks
(Anon 1936, p. 358).[25] However, as the decade progressed, the situation became
more difficult, particularly for Soviet scientists, the principal hindrances being the
increase in state control of the academic sphere, and the demand for 'ideological
orthodoxy': that all disciplines should remodel themselves in order to become con-
sistent with the Marxist philosophy of dialectical materialism. Science, in particular,
lay in the ideologues' sights, since its evidence-based nature made it attractive to
Marxist philosophers. However, as Alexei Kojevnikov has commented:

> despite their professed respect towards science, Bolsheviks with very few exceptions did
> not possess even basic scientific literacy and could be highly suspicious of scientists in real
> life. (Kojevnikov 2004, p. 280)

Nevertheless, this materialist orientation translated into an emphasis, though, it
seems, a largely rhetorical one, on experimental sciences.[26] Thus, as the 1930s
progressed, all academic disciplines, the sciences in particular, found themselves
under pressure to toe the ideological line, and we see many instances of ideological
interference in the sciences, the most infamous example being the influence of

---

[23] Korneyev and Timofeyev (1977, p. 10). Similar comments in the Soviet/American context can
be found in Furaev (1974). Such accusations continued to be made even until the final years of the
Soviet era; see Sapsai (1984) and Medvedev (1984).

[24] With regard to British-Soviet scientific relations, many more examples of exchanges may be
found in Korneyev and Timofeyev (1977). Other (slightly uncritical) reviews of British-Soviet
relations may be found in Topchiev (1956) and Romanovsky (1967).

[25] See also Tokareva (2001, pp. 219–222).

[26] On Marxist philosophy of science, see Graham (1972), Graham (1993, Chap. 5) and Todes and
Krementsov (2010); see also the summary in Gordin et al. (2003, pp. 39–43). In the case of mathematics, see Vucinich (1999, 2000, 2002) or Hollings (2013).

Trofim Lysenko over genetics—although, as has been demonstrated, ideology was in fact used here as a weapon in a fight over institutional resources (Joravsky 1970; Medvedev 1969). Indeed, it should be noted that it was not simply a matter of 'good' scientists versus 'bad' ideologues: many scientists took the Marxist viewpoint seriously, and used it to advance their discipline (Gordin 2014), whilst others used ideology for their own ends—young scientists in particular made ideological references in their published work in order to "bolster their appeals for state support" (Todes and Krementsov 2010, p. 354). We will see further cynical use of state ideology shortly, in connection with the so-called 'Luzin affair'.

As well as concerning themselves with the home-grown ideas of Soviet scientists, Marxist philosophers also worried about the import of 'idealistic' notions from the West. Increasing restrictions were therefore placed on the ability of Soviet scientists to communicate with their Western counterparts, lest they be 'corrupted', either politically or philosophically. Indeed, outside of some very narrow limits, international scientific communication was slowly and quite deliberately strangled; the activities of VOKS were disrupted (David-Fox 2002; David-Fox 2012, p. 91), particularly after the formation of a new body, Intourist (Ин[остранный]турист = Foreign tourist), to handle foreign visitors (David-Fox 2012, p. 175). VOKS and Intourist soon came into conflict, though both bodies were heavily affected by the subsequent purges—VOKS in particular was accused of having been too friendly in its dealings with foreigners, and too lax in its handling of foreign correspondence (David-Fox 2012, pp. 178, 193–194), and thus to have been 'tainted' by the 'inappropriate' foreign literature in its library (David-Fox 2012, p. 299). The flow of Soviet papers to Western scientific journals, which had until this point been steady and extensive (see below), largely dried up. Any scientist who still attempted to make contact with Western colleagues was regarded with suspicion, and was liable to find his- or herself accused of anything from the ideological sin of 'philosophical idealism' to the even more treasonable offence of being a counter-revolutionary.

One of the most high-profile episodes in this regard was the now-infamous 'Luzin affair' of 1936,[27] which saw the Moscow-based mathematician Nikolai Nikolaevich Luzin (Николай Николаевич Лузин) (1883–1950) summoned before a special commission of the Academy of Sciences on a range of spurious charges, including the accusation that he had passed off some of his students' work as his own, and that he had sought to 'undermine' Soviet science by publishing his best work in foreign journals. On the last count, Luzin's accusers probably felt that they had a wealth of evidence: of the 93 publications that are listed under Luzin's name in the 1959 survey volume *Mathematics in the USSR after Forty Years* (*Математика в СССР за сорок лет*) (Kurosh et al. 1959, vol. 2, 420–422),[28] 45 were published abroad. Moreover, he had extensive foreign contacts: he had spent

---

[27]There is an enormous literature on the 'Luzin affair'. See, for example, Demidov and Esakov (1999), Demidov and Levshin (1999), Kutateladze (2007, 2012, 2013), Levin (1990), Lorentz (2001, 2002), and Yushkevich (1989).

[28]For further details on this and other such survey volumes published in the USSR, see the references in note 37 on p. 68, and also Hollings (2015).

some time in Paris in 1905–1906 (Graham and Kantor 2009, pp. 80–82), and also as recently as 1928, the latter trip having been made thanks to Rockefeller funding (Siegmund-Schultze 2001, p. 295).

Almost inevitably, the Academy commission ruled against Luzin, and he was stripped of all his official positions, although the judgement against him was eventually overturned in January 2012 (Kutateladze 2013). Personal rivalries, and the possibility of career advancement for Luzin's accusers (who included several of his former students), probably had a role to play in this episode, as they did in other such attacks,[29] but the 'Luzin affair' nevertheless delivered a clear ideological message: that domestic publication was to be preferred. Indeed, although it did not initiate it, the 'Luzin affair' put the final seal on a trend towards domestic publication that had been underway since the beginning of the decade (Aleksandrov 1996). As an illustration of this tendency in the case of mathematics, we look again to *Mathematics in the USSR after Forty Years* and extract some data. The second volume of this survey work consists entirely of an impressive effort to list the publications, up to 1957, of every Soviet mathematician. As an illustration, let us select just three prominent figures (Luzin, P.S. Aleksandrov, and S.N. Bernstein[30]) and draw bar charts representing their levels of domestic and foreign publication up to 1957 (Figs. 2.1, 2.2 and 2.3).[31] As we see from the charts, all three had a history of extensive foreign publication prior to 1936, but then the figures drop off sharply after this date. However, we note also the fact that, as with many aspects of Soviet policy, the drive towards domestic publication was not applied consistently: foreign publication did not dry up completely—'ideologically sound' figures, such as P.S. Aleksandrov, remained able to send at least some of their work abroad.[32]

State suspicion of individuals with foreign contacts was not the only reason for the decline in foreign publication amongst Soviet scientists: nationalistic considerations also come into play. The 1930s saw a heightened

glorifying [of] elements of the Russian past [which] led to ignoring the achievements of non-Soviet scientists and to the isolation of Soviet sciences. (Lorentz 2002, p. 194)

---

[29] In connection with Luzin, see Graham and Kantor (2009, pp. 149–150), Kutateladze (2007), and Lorentz (2001). For another example of an 'ideological' attack which appears to have involved personal rivalries (namely, that on A.F. Ioffe in 1936), see Levin (1990, p. 97–8). The ambitions of younger researchers may also have played a role in the downfall of the geneticist N.I. Vavilov (Kolchinsky 2014). See also the case of the astronomer Boris Gerasimovich: Denny (1936).

[30] I take Aleksandrov as an example of a figure who was regarded as 'politically sound', and Bernstein as an individual who was more often at odds with the Soviet regime (see, for example, Vucinich 2000).

[31] Further data of this type, for some other Soviet mathematicians, may be found in Table 2.1 in Hollings (2014, p. 18), from which Figs. 2.1, 2.2 and 2.3 were also constructed. A similar, though slightly narrower, analysis appears in Levin (1990, p. 96).

[32] A *Pravda* article of 9th July 1936 (quoted by Lorentz 2001, p. 205) condemned Luzin's foreign publications as "sabotage", but counted those of Aleksandrov (amongst others) as mere lapses in judgement. Thus, we see that the issue of foreign publication was merely pretext in the 'Luzin affair'.

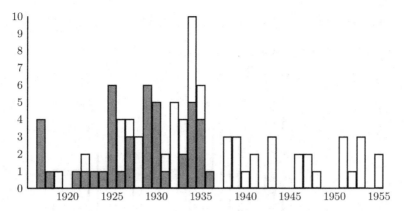

Fig. 2.1 Number of publications of N.N. Luzin per year, as listed in the volume Kurosh et al. (1959), showing both those items published domestically (*clear*) and abroad (*shaded*)

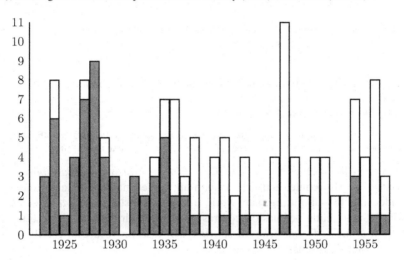

Fig. 2.2 Number of publications of P.S. Aleksandrov per year, as listed in the volume Kurosh et al. (1959), showing both those items published domestically (*clear*) and abroad (*shaded*)

In connection with publishing, this led to the feeling that the USSR ought to have its own world-class journals. To take another example from mathematics, let us consider an editorial entitled 'Soviet mathematicians, support your journal!' ('Советские математики, поддерживайте свой журнал!': Anon 1931) that appeared in volume 38 (1931) of the Moscow-based journal *Matematicheskii sbornik* (*Математический сборник = Mathematical Collection*). This editorial spoke out against the tradition, apparently prevalent amongst Soviet mathematicians,

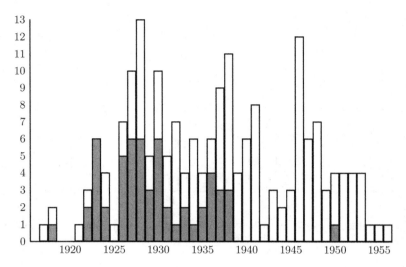

Fig. 2.3 Number of publications of S.N. Bernstein per year, as listed in the volume Kurosh et al. (1959), showing both those items published domestically (*clear*) and abroad (*shaded*)

"of publishing their best work in foreign journals",[33] and challenged the point of view that this was necessary to increase the visibility of Soviet mathematics around the globe; the editors opined instead that

> scattered throughout journals in Germany, France, Italy, America, Poland, and other bour-geois countries, Soviet mathematics does not appear as such, unable to show its own face.[34]

They thus set themselves the task of transforming *Matematicheskii sbornik* into a journal of world repute, and called upon all Soviet mathematicians to assist in this endeavour by submitting their work, in the first instance, to *Matematicheskii sbornik*.[35] Such gentle persuasion contrasts sharply with the less subtle effects of the 'Luzin affair'. *Matematicheskii sbornik* will also prove useful for illustrative purposes when we discuss languages in Chap. 4.

To return to the issue of foreign travel, we note that it remained possible for Soviet scientists to visit other countries in the early part of the 1930s—witness, for example, the three Soviet delegates at the 1932 Zurich ICM (Table 2.1). How-ever, as the decade advanced, the rights of Soviet citizens to travel were gradually

---

[33]"печатать свои лучшие работы в иностранных журналах" (Anon 1931).

[34]"рассыпанная по журналам Германии, Франции, Италии, Америки, Польши и других буржуазных стран советская математика не выступает как таковая, не может показать собственного лица" (Anon 1931).

[35]It is interesting to contrast this situation with that of several decades later: after the fall of the Soviet Union, Russian scientists came increasingly to the realisation that their work had a poor visibility at the international level. Their conclusion was that they ought therefore to return to the old tradition and begin once more to publish much of their work abroad (and, moreover, in English): see Kirchik et al. (2012).

curtailed, due not only to official fears that Soviet citizens would be 'corrupted' by foreign ideologies, but also, more simply, by the fact that some Soviet scientists who had travelled abroad had never returned (Josephson 1992, pp. 597–598)—the USSR sought to stem any further 'brain drain'. Thus, for example, of the eleven Soviet delegates who registered to attend the 1936 ICM in Oslo, none were in fact permitted to travel (Lehto 1998, p. 69). Indeed, in the second half of the 1930s, Soviet scientists became quite conspicuous by their absence from international sci-entific meetings: as disparate examples, we might take the Second International Forestry Congress (Budapest, 1936: see Guthrie 1936), the Warsaw Conference on Modern Physics (1938: see Anon 1938), and the Seventh International Genetics Congress (Edinburgh, 1939: see Krementsov 2005, p. 4, or Crew 1939), none of which included any Soviet delegates; tentative plans had been made to hold the last conference on this list in Moscow in 1937, but political difficulties had caused these to fall through. Thus, Soviet scientists found themselves in what Kojevnikov has referred to as

> the twenty years of Stalin's dictatorship, when Soviet science worked in virtual international isolation, with practically no foreign travel, visits, personal communications, conferences, or correspondence, and when most contacts with the rest of the world science would be reduced to exchanges of printed works. (Kojevnikov 2004, p. 85)

Such exchanges of printed works will be considered in Chap. 3.

Although the 1930s represent a dark period in Soviet history, hardly ideal for fos-tering a thriving international scientific community, there was the occasional glim-mer of hope. Soviet researchers may not have been able to travel abroad, but foreign scientists were still able to visit the USSR, though perhaps not as easily as they had been able to do so earlier: in contrast to the above-mentioned First All-Union Mathematical Congress of 1930, the Second All-Union Mathematical Congress (Leningrad, 1934) hosted just a single foreign delegate (Anon 1935a, vol. 1, p. 14).[36] From 1938, VOKS ceased to invite and host foreign visitors, concentrating instead on cultural activities abroad (David-Fox 2012, p. 305).

Many of the foreign scientific visitors to the Soviet Union during the 1930s were there for practical reasons. For example, a large number of technical specialists were invited to the USSR in order to assist in building up Soviet industry (Medvedev 1979, p. 28); according to some sources, there were, by 1932, 600 American engi-neers working in Soviet car and tractor plants (Byrnes 1976, p. 29; Kuznick 1987, pp. 113–114). Much as it had in the preceding decade, VOKS also hosted delega-tions of foreign scientists (at least until 1938), such as that which visited the Soviet Union in the Summer of 1931, and which included, amongst others, the biologist Julian Huxley[37]; the aim of issuing such invitations was, one would assume, to impress visitors with the rapid technical achievements of the USSR. Some indi-vidual Western scientists were also permitted to work in Russian universities: for example, the American geneticist H.J. Muller, who spent time in both Moscow and

---

[36]See also the report of that (American) delegate: Lefschetz (1934).

[37]See Huxley (1932); see also Kuznick (1987, p. 118).

Leningrad.[38] Finally, we note that the occasional international conference was held in the USSR during this period: indeed,

> [t]he Soviet government lavishly funded each of these gatherings, and the Soviet press covered them at every turn.[39]

Plans for the above-mentioned genetics congress may have fallen through in 1937, but the USSR had nevertheless played host to the Seventeenth International Geological Congress in that year (in conjunction with a meeting of the International Paleontological Union),[40] and to the Fifteenth International Physiological Congress just a couple of years earlier,[41] as well as two international mathematical congresses around the same time: one on vector and tensor analysis in 1934 (Anon 1934, p. 648), and another on topology in 1935 (G.B 1935).. Details on the former are lacking, but it does appear to have attracted delegates from Austria, Czechoslovakia, France, Germany, Italy, the Netherlands, and Poland. The topological congress, on the other hand, was a much larger affair; it has been described as "the first truly international conference in a specialized part of mathematics" (Whitney 1989, p. 97). The proceedings of this congress were published in volume 1(43) (1936) of *Matematicheskii sbornik*, from which we see that delegates hailed from Czechoslovakia, Denmark, France, Germany, the Netherlands, Norway, Poland, Switzerland, and the USA.[42]

2.4 The Second World War

The partnership between the USSR and the other Allied nations during the Second World War, at least after the Soviet Union's entry into the conflict in 1941, was attended, as one might expect, by an enhanced spirit of cooperation.[43] Efforts to connect with, and, indeed, to understand, the peoples on the opposite side of what was to become the Iron Curtain were stepped up in all walks of life, not least in the sciences. However, many such efforts appear to have been offset by new barriers to intercourse: not now (necessarily) the difficulties created by Soviet policy, but the more general problems of wartime communication. Thus, despite a cooperative desire on both sides to exchange material, little contact of practical value appears to have been achieved.[44] Instead, we find many rhetorical statements of solidarity

---

[38] See Kuznick (1987, pp. 119–125), and also Carlson (1981, 2011).

[39] Doel et al. (2005, p. 59). See this source also for other examples of international congresses held in the USSR during the 1930s.

[40] See Krementsov (2005, p. 8), and also Gordon (1937) and Case et al. (1938).

[41] See Franklin (1938, pp. 314–320) and Kuznick (1987, pp.153–162); see also Anon (1935b) and Ivy (1935).

[42] For reports of the congress from both sides of the East/West divide, see Aleksandrov (1936) and Tucker (1935).

[43] Indeed, prior to this, there had been greater cultural ties between the USSR and Nazi Germany, during the years of the Molotov–Ribbentrop Pact; see David-Fox (2012, p. 310–311).

[44] I am confining my attention here to 'civilian science'—the type of science found in freely published papers, and also the type that we have been concerned with implicitly from the start of this

and cooperation, such as that signed by 93 American mathematicians in 1941, and delivered to the Soviet embassy in Washington. In this message, which was subsequently printed in both *Science* and *Nature*, the mathematicians sent their "greetings and . . . heartfelt sympathy to [their] colleagues of the Soviet Union in their struggle against Hitler fascism [*sic*]" (Anon 1941a), before going on to remark that

> [t]he bonds between mathematicians in the United States and the Soviet Union are particularly strong since during the past two decades the center of world mathematics has steadily shifted to these two countries. We know many of you personally and more of you through your scientific writings. (Anon 1941a)

The mention of 'particularly strong' bonds here is perhaps a little questionable. Nevertheless, we find the same high-flown language in a response, signed by 64 Soviet mathematicians, sent some weeks later:

> Your splendid message, dear colleagues, found wide response in the hearts of the scientists of our country. We read it with feelings of all the more appreciation and satisfaction in that it again emphasized the community of thought and the friendly ties between the mathematicians of the U.S.A. and the U.S.S.R. Many years we jointly worked with you on the development of our science, many of our American colleagues were our welcomed guests, while with a still greater number of American scientists we conduct friendly scientific correspondence. This mutual co-operation was very fruitful and led to a number of important scientific discoveries. (Anon 1941b)

Similar statements of wartime scientific solidarity were exchanged, for example, by the Linnean Society and the Moscow Society of Naturalists (Anon 1942f), by the American Association of Scientific Workers and the Soviet Scientists' Antifascist Committee (Anon 1943a), and by the Royal Society and the Soviet Academy of Sciences.[45] In connection with the latter, hopes were raised that the exchange of scientific information might be improved, but it is unclear whether anything of practical value was ever achieved in this regard.[46] For further expressions of solidarity, we may also look to the comments concerning British and Soviet scientists that appear in the proceedings of a symposium on Soviet science that was held at Marx House in London at Easter 1942.[47] However, since this conference was held under the auspices of a Marxist organisation, these remarks are of a rather more political and propagandistic tenor than those found in other places; the proceedings also include (in English) appeals made by the Soviet Academy of Sciences which exhort foreign colleagues to aid in the fight to "wipe the brown pestilence of Fascism from the face of the earth" (Anon 1942h, p. 31).

Communications of a slightly more practical nature did occasionally slip through the surrounding rhetoric. These included, for example, book-length surveys of Soviet science, which, unsurprisingly, focused on those aspects of Russian work

---

(Footnote 44 continued)
book. The exchange of military technology and of secret materials is not something that I attempt to cover here—see instead, for example, Beardsley (1977) and Avery (1993).

[45]There is a wide range of letters to cite in this instance: see, for example, Anon (1941c, 1942d, 1943c); see also Korneyev and Timofeyev (1977, p. 34).

[46]See Anon (1942a, e, g).

[47]See Anon (1942h), and also Anon (1942b, c).

that were then aiding the war effort.[48] In addition, we find, in Western general science journals, various survey articles, by both Western and Soviet authors, detailing the current status of, or recent progress in, specific disciplines within the Soviet Union: take, for example, those on mathematics (Vinogradov 1942), astronomy (Anon 1943b), botany (Shishkin 1943), biology (Dunn 1944), and physics (Joffe 1945), as well as a considerably longer one covering chemistry, physics, metallurgy, radio telegraphy, and aeronautics (Ipatieff 1943). In the opposite direction, the Soviet Academy of Sciences published summaries of Western (mostly British and American) scientific works (Krementsov 1996, p. 234).

Around this time, communications of a similar type were taking place also via the British-sponsored journals *Британский союзник* (*Britanskii soyuznik* = *British Ally*) and *Британская хроника* (*Britanskaya khronika* = *British Chronicle*), which had been founded in the wake of the Soviet-British Treaty of 1942 to provide the Soviet people with a window onto British life. As an uncensored account of a part of the world beyond the Soviet border, *Britanskii soyuznik* proved very popular within the USSR, and eventually inspired a glossier, though more overtly propagandistic, American counterpart *Америка* (*Amerika*) (Byrnes 1976, pp. 30–31). However, although these journals covered a range of subjects, including science and technology, their contents, which focus on how science was aiding the war effort,[49] appear to have been too superficial to have contributed much to international scientific communication.[50] The same can also be said of *Britanskii soyuznik*'s counterpart, *Soviet War News*, which was published daily by the Press Department of the Soviet embassy in London between 1941 and 1945. Again, this newspaper sought to inform the British public about Soviet life in general, and was taken up very largely by the requisite expressions of solidarity, and by accounts of how the Soviet people (including scientists[51]) were working to combat the fascist threat. Perhaps as a reflection of the supposed importance of science to Soviet thinking, *Soviet War News* appears to have featured more articles of a scientific nature than *Britanskii soyuznik*, but they were no less superficial.[52] Thus, although they contained very little scientific information, these journals, along with the sporadic survey articles mentioned above, must have gone at least some way towards informing the scientists on one side of the divide about what those on the other side were doing, even if they did little to increase the scope for personal communication between scientists in East and West.

It should be noted, however, that there was one group of scientists who did in fact manage to set up strong lines of communications between East and West during the war, namely those working in medicine. Cooperation in medical matters was of course born of the importance of such researches to the war effort, and also perhaps of a Western perception that the Soviet Union was making great strides in

---

[48] See, for example, Needham and Davies (1942) or Anon (1944).

[49] See, for example, Bernal (1944) and de Andrada (1944).

[50] See Johnston (2011, pp. 86–87) or Pechatnov (1998).

[51] See, for example, Anon (1941d, f, h, i), and Frumkin (1941).

[52] See, for example, Anon (1941e, 1945e), Rostov (1945), and Zhukovsky (1945).

medical matters[53]; the result was the exchange of medical research much more generally. Upon the USSR's entry into the war, an Anglo-Soviet Medical Committee was formed in London, with the goal of obtaining information on Soviet medical techniques, and of sending similar details about British advances to the USSR.[54] The Soviet ambassador was approached in connection with the former, and appeals went out for volunteers, not only to translate Russian medical materials into English, but also to prepare a Russian translation of a volume entitled *Reviews of British War Medicine*; upon completion, the latter translation was presented to the wife of the Soviet ambassador in November 1942 (Anon 1942j). Copies of *The Lancet* and of *The British Medical Journal* were also dispatched to Russia (Anon 1941k), along with issues of the wartime journal *Bulletin of War Medicine*; copies of Russian materials were received in turn. Other activities of the Anglo-Soviet Medical Committee included the sending of spare surgical equipment to the USSR (Webb-Johnson 1941), and the facilitation of exchanges of British and Soviet medical personnel; reports of trips to the USSR, featuring details of Soviet medical advances, thus began to appear in British (more generally, Western) publications[55]—such reports (in a more general setting) will feature again in Sect. 3.5.

Across the Atlantic, perhaps following the British example, an American-Soviet Medical Society emerged in 1943.[56] The goals of the American society were broadly similar to those of the British one: to promote medical links with the USSR, and to facilitate the exchange both of printed materials and of personnel—evidence that this did indeed take place on a significant scale is found in the number of foreign letters processed by a revived VOKS (Krementsov 1996, p. 233–234). In connection with information exchange, the society built up an extensive library of Soviet medical texts (Lear 1997, p. 270), whilst in the case of exchange of personnel, it arranged lectures by US medical researchers who had just returned from the USSR, and also by Soviet visitors (Lear 1997, pp. 270–274). The main activity of the American-Soviet Medical Society, however, was the publication of a journal, *The American Review of Soviet Medicine*, which carried English translations of major Russian articles, and abstracts of others, along with surveys of specific branches of Soviet medicine.[57] A foreword to the first issue noted that "[t]he medical profession is the world's greatest fraternity", and proudly asserted that "[i]n medical research, ... no artificial barriers between nations are recognized" (Cannon 1943, p. 5). Moreover, an editorial in the same issue expressed the hope "that [*The Review*] will become a permanent link between the medical corps of our two great countries" (H.E.S 1943). Indeed, throughout the remaining years of the war, *The American Review of Soviet Medicine* continued to be a steady connection between the medical researchers of the USSR and the USA, and in the few years following the war, the American-Soviet

---

[53] See Carling (1944) or Lear (1997, pp. 259–260).

[54] See Anon (1941g, j, 1942i), Dawson et al. (1941a, b), and Bunbury (1942).

[55] See, for example, Watson-Jones (1943a, b) and Hastings and Shimkin (1946). On UK-USSR medical exchanges in later decades, see Rich (1975).

[56] See Lear (1997) or Krementsov (2007, pp. 44–48).

[57] See Lear (1997, pp. 267–270) or Kerber (2012).

Medical Society aided in the restocking of Soviet medical libraries by sending regular shipments of American materials (Lear 1997, p. 271), but its activities soon petered out in an unfavourable post-war climate.

2.5   After the War

As is well documented, the spirit of wartime unity between the USSR and the other Allies quickly dissolved in the wake of the defeat of Nazi Germany, the Iron Curtain descended, and the path into the Cold War was set. Scientific contacts during the war, if somewhat limited, had nevertheless led many scientists to believe that a new era of post-war cooperation would begin, but it was not to be. This was due, in very large measure, to a return to the form of the 1930s in connection with Soviet policy. As Loren Graham has commented:

> [a]fter the Second World War many intellectuals in the Soviet Union hoped for a relaxation of the system of controls that had been developed during the strenuous industrialization and military mobilizations. Instead, there followed the darkest period of state interference in artistic and scientific realms. (Graham 1972, p. 18)

Indeed, Soviet scientists were not alone in their initial hopes that international contacts might be strengthened: efforts were made by US agencies not only to establish a student exchange programme between the USA and the USSR, but also, in 1945, to instigate a Rhodes-type scholarship (Byrnes 1976, pp. 31–33)— educational exchange generally, and that of science education in particular, was seen by some as a route to international understanding.[58] These endeavours, however, came to nought, for this was the era of *zhdanovshchina* (*ждановщина*)— named for Stalin's chief ideologist A.A. Zhdanov, this was the post-war Soviet policy whereby the Western influences that had crept into Soviet life during wartime cooperation were to be purged, and 'cultural purity' was to be promoted. In this climate the journals *Britanskii soyuznik* and *Amerika* came to be regarded by the Soviet authorities with greater suspicion, and were lumped together with the more blatant Western (particularly American) propaganda. Moreover, the fact that both journals were often mistaken by the Soviet populace for domestic publications leant them an insidious air. Their days were numbered when V.S. Abakumov, the Soviet Minister of State Security, wrote to both Stalin and Zhdanov to express his concerns (Levering 2002, pp. 165–166). Both journals had ceased publication by the end of the 1940s. The rapidly descending Iron Curtain also impacted Western publications: the renewed difficulties of obtaining Russian manuscripts, coupled with the increasing anti-communist feeling in the USA, meant that the October 1948 issue of *The American Review of Soviet Medicine* was its last.[59] It should be noted, however, that the cessation of the publication of *The American Review of Soviet Medi-*

---

[58] See, for example, Oakes (1946) and Bu (1999).

[59] See Lear (1997, pp. 274–276), Sigerist (1948) or Anon (1948). Interestingly, however, other similar journals were being launched elsewhere around this time: the Spanish-language

cine, and, more generally, of the activities of the American-Soviet Medical Society, was not due solely to political pressures from above, but also to a dwindling interest on the part of former subscribers: an editorial in one of the last issues of the journal expressed disappointment at the waning enthusiasm for the journal and for the society amongst US doctors now that the USSR was no longer a comrade-at-arms and was in fact becoming increasingly unpopular in the American press (Sigerist 1948, p. 7).

Immediate post-war prospects for renewed scientific exchange between East and West had in fact seemed quite promising, with the invitation of many foreign delegates (apparently at Stalin's suggestion: see Krementsov 2007, p. 46) to the celebration of the 220th anniversary of the Russian Academy of Sciences in Moscow and Leningrad in June 1945,[60] and the appearance of small Soviet delegations at the conference 'Science in Peace', held in London in the February of the same year (Anon 1945c), at the Royal Society's Newton Tercentenary Celebrations in July 1946,[61] and at the Seventeenth International Congress of Physiological Sciences in Oxford in 1947 (Fenn et al. 1968, pp. 24–30). However, events were soon overtaken by the principles of 'zhdanovshchina', and, in 1947, a new Soviet law was passed, which decreed that no individual or organisation could make contact with foreigners without the express permission either of the Ministry of Foreign Affairs or of the Ministry of Foreign Trade (Byrnes 1976, p. 32). Moreover, Soviet delegations sent to foreign conferences were given strict instructions, handed down directly from the Central Committee, on how exactly to behave (Krementsov 2007, pp. 60–61). International scientific contacts were thus by no means impossible, but they certainly remained difficult. Even those scientists who were permitted to publish some of their work abroad (in particular, in the West) found themselves under attack: an article in *Pravda* in mid-1947 criticised a number of such scientists for their supposed 'unpatriotic acts' and 'servility to the West' (Gerovitch 2002, p. 15). The infamous 'honour trial' of the biomedical researchers G.I. Roskin and N.G. Klyueva, at which identical accusations were made, also took place in 1947 (Krementsov 2002, pp. 109–133). Another example is provided by the attack on the geneticist A.R. Zhebrak, which took place the same year, and again centred upon accusations of 'servility to the West'.[62] During this period, ideology once again became the basis for assaults on science and scientists; take, for example, the group of Leningrad-

---

(Footnote 59 continued)

Revista cubana de medicina sovietica, for example, was founded in 1945, whilst the French *Cahiers de médicine soviétique* ran from 1953 to 1957; see Kerber (2012, pp. 233–234). The French journal can perhaps be seen as a successor to an earlier Soviet-French medical copublication: see Ivanovskaya (1976, pp. 201–202).

[60] See Krementsov (1996, p. 237), Krementsov (2002, pp. 75–78) or Anon (1945a, b, d); see also Korneyev and Timofeyev (1977, pp. 38–39). *Soviet War News* had much to say about this conference, its international character in particular; see, for example, Anon (1945f, g, h, i).

[61] See Dale (1946, p. 157); see also Korneyev and Timofeyev (1977, p. 40).

[62] See Krementsov (2005, p. 142). Zhebrak's publication of an article in *Science* (Zhebrak 1945) was, for example, held against him. See also the comments of Medvedev (1979, p. 119), not only on Zhebrak, but also on Roskin and Klyueva.

based mathematicians who came under fire in 1949 for their supposedly 'idealistic' research pursuits.[63]

By and large, the Western follow-up to wartime contacts, at least during the late 1940s, could not have been more different from the official Soviet line: sitting in contrast to Stalinist concerns about 'cultural contamination' was a piqued Western curiosity about all things Soviet—or at least an uneasy feeling that Westerners ought to know more about the USSR. In its early stages, this renewed interest manifested itself in concerns about the provision of resources for academic research in Soviet, Russian, Slavonic and/or Eastern European studies: witness for example the report commissioned by the UK's Foreign Office, in which we find the following statement of motivation:

> The comradeship of the war and the supreme importance of continuing that comradeship in the future furnish the strongest reasons for developing in Great Britain sound and accurate knowledge of the Soviet Union and the Russian way of life. In the inter-war period political conditions were unfavourable to the spread of accurate knowledge about the Soviet Union and there is much leeway to be made up. (Foreign Office 1947, p. 26)

The report's authors asserted that there was then

> clear evidence of a strong desire in [the UK] to learn more about the Soviet Union and we have been informed that a corresponding desire exists in the Soviet Union and that much attention is given in that country to the study of Great Britain and the British Empire. (Foreign Office 1947, p. 26)

The report thus set out to survey the then-current status of Russian studies within British universities, and to make recommendations for improvement. We find therein, for example, a recognition of the fact that "[o]f the large amount of scientific work" produced by the Soviet Union (amongst other countries),

> a substantial proportion is not available to scientists in [the UK] on account of the barrier of language. (Foreign Office 1947, p. 33)

The provision of more translators is therefore one of the report's recommendations. The issue of language is one to which we will return in Chap. 4. Concern over Slavonic and related studies, including the language problem, was also felt elsewhere in Western Europe: in France, for example (Mazon 1946; Hilton 1979).

I have not yet come across any report of similar scope relating to US academia,[64] but one American reviewer of the British report opined that its findings were "no less true for the United States" (Frye 1947, p. 333)—that general US understanding of the USSR was not yet adequate. Indeed, this appears to have been recognised in the United States as early as 1941, for this was the year in which the journal *The Russian Review* was launched; in the foreword to the first issue, we find the sentiment that

---

[63] See Gerovitch (2002, pp. 34–35) and Hollings (2012).

[64] The closest I have come is a state-by-state directory of the Russian culture and language courses offered by US higher educational institutions: Coleman (1948). See also Strakhovsky (1947). Much more generally, a US State Department report of 1950, *Science and Foreign Relations*, stressed that awareness of foreign scientific developments was crucial to the progress of US science; see Krige (2006, p. 166).

Russia is much less known to Americans than its size, its political importance, and its contributions to culture would warrant. (Chamberlin 1941, p. 1)

The journal continues to this day, with one online archive describing it as

a multi-disciplinary academic journal devoted to the history, literature, culture, fine arts,
cinema, society, and politics of the peoples of the former Russian Empire and former Soviet
Union.[65]

Historically, the journal has also carried articles on Soviet science, some of which
are cited in the present book. Over the following decades, *The Russian Review* was
joined by a number of other journals devoted specifically to Soviet culture—I will
say a little about these in Sect. 3.5, in which we will consider, more generally,
the many surveys of Soviet science that were subsequently produced for Western
readers.

2.6   The Post-Stalin Period

Stalin's death in 1953 led to dramatic changes in almost all aspects of Soviet life, as
the (slightly) more liberal atmosphere of Khrushchev's 'thaw' took effect. Despite
some notable exceptions (Lysenkoism in particular), the application of state ideology to the sciences became rather less dogmatic. Certainly, those sciences in which
Soviet researchers enjoyed a prominent presence on the world scientific stage (mathematics, for example[66]) found themselves in a much more secure position: patriotic
pride outweighed philosophical considerations. Nevertheless, state ideology had not
gone away, and Soviet scientists (those who did not attempt to put ideology to cynical use) were required at least to acknowledge it.[67] Others were able to employ
ideological language for their own ends.[68]

Khrushchev's thaw also introduced more scope for international scientific communication, with the replacement of the now largely defunct VOKS by the new
Union of Soviet Societies of Friendship and Cultural Relations with Foreign
Countries   (Союз советских обществ дружбы и культурных связей с
зарубежными странами) (Smith 2012, p. 548). Slava Gerovitch has commented
that

Soviet scholars could now publish abroad, attend international conferences, receive foreign literature, and invite their foreign colleagues to visit. The division into 'socialist' and
'capitalist' science no longer held; claims were made for the universality of science across
political borders. (Gerovitch 2002, p. 155)

---

[65]http://www.jstor.org/action/showPublication?journalCode=russianreview (last accessed 26th
May 2015).

[66]On the international standing of Soviet mathematics, see Graham (1993, pp. 213–220) or
Dalmedico (1997).

[67]In the case of mathematics, for example, see Vucinich (2002).

[68]Indeed, this matter is the main theme of Gerovitch (2002).

In some regards, the Soviet authorities may even have sought to *encourage* international scientific cooperation, probably with a view to catching up with the West in those disciplines in which Soviet research was perceived to lag behind (Medvedev 1979, Chap. 6). Some commentators have indeed given this as one of the USSR's major motivations for engaging in academic exchanges, citing the prominence of Soviet scientists in the cultural exchanges that followed Stalin's death (Byrnes 1976, p. 73). In this connection, Gerovitch has written on the shift in the official Soviet attitude towards Western science from 'criticise and destroy' to 'overtake and surpass' (Gerovitch 2002, pp. 18–21, Chap. 4), and has remarked further upon the

> detailed instructions [that were issued by the Presidium of the Academy of Sciences] on how to obtain the permission for a foreign trip, how to invite foreign colleagues, how to obtain the permission to publish an article abroad, and how to maintain correspondence with foreign scholars and scientific institutions. Restrictive as they were, these instructions nevertheless legitimized what had been unthinkable in the late Stalinist period: regular contacts and exchanges between Soviet scientists and their Western colleagues. (Gerovitch 2002, pp. 156–157)

Nevertheless, although procedures were in place to enable international communication, these did not always run smoothly, as we shall see: the Soviet physicist R.Z. Sagdeev noted that, in terms of the ease of obtaining it, "permission to take foreign trips was almost a ticket to outer space" (Sagdeev 1994, p. 137).

One of the first formal agreements on academic exchange between a Western nation and the USSR was that devised in 1959 by the American Council of Learned Societies and the Soviet Academy of Sciences.[69] The first exchange under this agreement took place in 1961 when a group of four US academics (a Pushkin scholar, an archaeologist, an economist, and a historian) travelled to the USSR; a reciprocal visit by three Soviet economists and a historian was arranged the following year (Anon 1962a). In the decades that followed, many hundreds of scientists travelled between the United States and the Soviet Union, and vice versa, sponsored by their respective academies of sciences (Schweitzer 1989, p. 171). The exchange agreement was continually modified and added to over the years,[70] and gave the opportunity, for example, for US students to enrol in Soviet universities, and for Soviet professors to lecture in the USA (Schweitzer 2004, p. 2). Indeed, the establishment of procedures for formal collaboration enabled exchanges to take place in disciplines which had hitherto seen little peacetime East-West cooperation, such as medicine.[71] From the late 1950s onwards, the general scientific literature contains many examples of Western scientists visiting the USSR, and of Soviet scientists making the opposite trip; a particularly historic example of the latter was the visit to the USA made by four non-Lysenkoist Soviet geneticists, two years after Lysenko's removal from the Soviet Institute of Genetics (Langer 1967). Moreover, science-

---

[69] See Schweitzer (2004, Chap. 1); the text of the formal agreement is reproduced in Schweitzer's Appendix B.

[70] See, for example, Anon (1963a, 1972a).

[71] On US-USSR cooperation in medicine, see Raymond (1973); on UK-USSR cooperation, see Rich (1974).

related scholarly exchanges were not merely the preserve of 'pure' scientists: witness, for example, a visit to the USSR by a delegation from the Federation of British Industries (Anon 1963c), and also that undertaken by a group of science librarians (Francis 1963).

Aside from the altruistic motivation of cooperating with 'the other side' in order to contribute to scientific advancement, other reasons for academic exchange were at play. As I have already indicated, the Soviet authorities had a view to using these exchanges to catch up with the West, most particularly in technological terms. Indeed, the USSR did, on occasion, attempt to abuse the exchange programmes for military gain (Schweitzer 1989, p. 194). The chance to earn recognition on the world scientific stage may also have played a part (Schweitzer 2004, p. vii). On the other hand, remaining fears about 'philosophical contamination' were joined by renewed concerns that free travel to the West might result in the 'brain drain' of Soviet science.[72]

In the USA, the possibility of scientific exchange with the USSR was not always greeted with enthusiasm (Byrnes 1976, p. 74), possibly because it was felt by some that Soviet science had nothing to offer—we will encounter this attitude again in Sect. 4.5 in connection with the translation of Soviet scientific works. There was a fear that Soviet scientific visitors were simply

> going round the United States like vacuum cleaners sucking up all kinds of scientific information and technical know-how, (Richmond 2003, p. 66)

and offering little in return. In general, however, a more open attitude to exchange prevailed, perhaps since such contacts were seen by many not only as a means of acquiring information, but also as a starting point for greater cultural contacts, since it was felt that

> [a]s a group, the scientists of the USSR are more open to considering ideas from abroad than are many other segments of the society. (Schweitzer 1989, p. 145)

A rather more blunt way of expressing US intentions would be to say that they were political and ideological in nature:

> to develop, within the framework of détente, patterns of cooperation and interdependence that would lead to shared interests and more moderate behavior on the part of the Soviet Union. (Richmond 2003, p. 69)

Indeed, as in earlier decades (see Sect. 2.2), scientific exchange served as a stealthy route to increased US influence—the view under the Eisenhower administration (1953–1960) was that

> increased international collaboration would strengthen the Free World, ease Cold War tensions, and promote the growth of science.[73]

---

[72] See Schweitzer (2004, p. 7) or Medvedev (1971, pp. 155–161).

[73] Hamblin (2000a, p. 393). On this subject, see also Wolfe (2013). For similar attitudes in a different context, see Wang (1999). On the use of science in US foreign policy more generally, see Miller (2006).

In a slightly broader setting, NATO also sought to strengthen Western science by promoting international collaboration (Krige 2000). The ultimate effects of such 'back door diplomacy' may indeed have been dramatic—see Sect. 2.9. It should be noted, however, that people in the USSR were not blind to US intentions: the Soviet 'Americanologist' G.A. Arbatov noted in 1969 that "[u]nderlying U.S. policy is the so-called 'erosion' of our social system".[74] Arbatov was, in general, in favour of cultural exchanges for the benefits that they would bring, but warned that the USSR should not lose sight of the USA's motivations. Indeed, it was perhaps in recognition of ulterior motives that the USSR had stopped accepting Rockefeller funding in 1933 (Krementsov 2005, p. 7, 42).

Viewed as foreign aid, scientific exchange became something that the USA could use as a bargaining chip, to be withheld until progress was made in other areas, such as human rights.[75] Exchange programmes also broke down at times for political reasons: US-Soviet scientific collaboration experienced a lull in 1967–8, for example, owing to various factors, including the ongoing Vietnam War and the Soviet invasion of Czechoslovakia (Byrnes 1976, pp. 47–48). Nevertheless, the decades following Stalin's death saw the organisation of exchanges and bilateral workshops in a vast range of areas (Schweitzer 1989, p. 165). Some small amount of cooperation even took place in space.[76]

Naturally, the United States was not the only Western nation to enjoy scientific exchanges with the USSR. Canada, Australia, France, West Germany, Belgium and Norway certainly had exchange programmes,[77] as did the UK, where the Royal Society organised exchanges with the Soviet Academy of Sciences.[78] Nuclear energy appears to have featured heavily in Soviet-French exchanges,[79] although other areas were also represented.[80] With regard to Soviet-British scientific relations, we have the volume, Korneyev (1977), mentioned in Sect. 2.2, which, if one looks past the melodramatic language that is sometimes used ("the imperialist forces have on many occasions attempted to stifle the Soviet state": Korneyev and Timofeyev 1977, p. 9), does appear to be factually accurate where specific examples of scientific exchange are given—both the 1925 and 1945 Academy of Sciences anniversary conferences appear, for example. Elsewhere (Korneyev 1977, pp. 295–319), we find samples of the "vast and varied" correspondence between British and

---

[74]Quoted in Richmond (2003, p. 18).

[75]See Schweitzer (2004, p. 8). Indeed, such behaviour was nothing new: German scientists had attempted to exert some influence over the trials of suspected German terrorists in the USSR by threatening to boycott the 1925 conference to celebrate the 200th anniversary of the Russian Academy of Sciences; see Forman (1973, p. 168). In the later US-Soviet context, the treatment of both dissidents and refusenik scientists became a major reason for postponing exchanges; see, for example, Rich (1979), Anon (1980), and Lubrano (1981).

[76]See Greenberg (1962) and Harvey and Ciccoritti (1974). On cooperation in later decades, see Edelson and Townsend (1989).

[77]See Byrnes (1976, p. 64), Richmond (2003, p. 15) or Nygren (1980).

[78]See Smith (2012, p. 550); see also the references in note 24 on p. 14.

[79]See, for example, Zavyalskii (1973), Kirillov (1977), Semenov (1979), and Isaev (1979).

[80]See, for example, Novikov (1973) and Aver'yanov and Korotkevich (1978).

Soviet scientists during the years 1955–1961, which, in the rather curious phrase of the book, "followed the decline of the 'cold war' " (Korneyev 1977, p. 295). One particular criticism to make of this volume, however, is that it presents the growth in scientific communication between the UK and the USSR as a uniform trend, rather than the halting process that other sources show it to be.

In this period, Soviet scientists were again beginning to appear at international conferences. We see from Table 2.1, for example, that, after an absence of roughly 20 years, Soviet mathematicians began once again to attend the ICMs, starting with the Amsterdam congress of 1954. Also in 1954, five Soviet delegates appeared at the Fourteenth International Congress of Psychology in Montreal, after a similarly long absence (Rosenzweig 2000, p. 74). In the following year, the USSR sent the third largest delegation (after the USA's and the UK's) to the Geneva International Conference on the Peaceful Uses of Atomic Energy,[81] whilst around 20 Soviet astronomers attended the Ninth General Assembly of the International Astronomical Union (IAU) in Dublin (Redman 1955), although, curiously, Soviet participation in the activities of the IAU do not appear to have dried up to quite the same extent as those in other disciplines. Indeed, for many years, the IAU was the only international scientific organisation of which the USSR was a member, having joined in 1935.[82] In contrast, the Soviet Union did not, for example, join the International Union of Physiological Sciences until 1953 (Fenn et al. 1968, p. 99), the ICSU (p. 11) until 1954 (Lehto 1998, p. 123), or the International Mathematical Union until 1957 (Lehto 1998, p. 122). With regard to astronomy, Soviet delegates to the IAU were proposing, as early as 1946, to stage a General Assembly in the USSR. However, concerns about official Soviet ideological condemnations of Western science, amongst other factors, prompted the IAU's executive committee to decline several such invitations in the late 1940s and early 1950s.[83] Nevertheless, several foreign astronomers were able to attend the reopening of the Pulkovo Observatory near Leningrad in 1954 (Anon 1954), and the IAU eventually held its Tenth General Assembly in Moscow in 1958, which was attended by delegates from 38 different countries (Blaauw 1994, Sects. 8.l–8.m). Conversely, a large Soviet delegation was able to attend the Eleventh General Assembly in Berkeley, California, three years later (Blaauw 1994, Sect. 10.d). To turn to other disciplines, we find, for example, that a number of foreign delegates attended the Moscow conferences on high-energy physics in 1956 (Pickavance and Skyrme 1956), and on oncology in 1962,[84] whilst Soviet delegates were present at a range of international scientific conferences throughout the 1950s, particularly in the second half: take, for instance, those on the effects of nuclear weapons (London, 1955), genetics (Tokyo and Kyoto, 1956), astronautics (Rome, 1956), radioisotopes (Paris, 1957), oceanography (New York

---

[81] See Krige (2006, pp. 174–180) or Schroeder-Gudehus (2012, pp. 31–32).

[82] See Struve (1953) or Blaauw (1994, p. 113).

[83] See Blaauw (1994, Sect. 8.d) or Doel et al. (2005, p. 67).

[84] See Krementsov (2002, p. 204) or Anon (1962b).

City, 1959), and physiology (Buenos Aires, 1959).[85] Moreover, this trend continued into the 1960s, with Soviet delegates in attendance of international conferences on space science (Washington, DC, 1962), biochemistry (New York City, 1964), and physiology (Tokyo, 1964), to name but a few.[86] One complication that should be noted, however, in connection with Soviet attendance of foreign conferences was the occasional insistence by Soviet delegates upon delivering their lectures in Russian, necessitating the use of an interpreter, even when the speaker was fluent in a more widely understood language, such as German, French or English.[87] The language issue will be treated in more detail in Chap. 4.

Thus, as the 1960s advanced, scientific exchanges between East and West were certainly on the increase. However, these were not without their difficulties, particularly on the Soviet side. To put things in a nutshell, the procedures established by the USSR supposedly to enable its scholars to communicate with, or even to travel to meet, counterparts of other nations, were generally hindered, often quite severely, by bureaucracy, and by the cynical use of bureaucracy. A quite comprehensive treatment of the difficulties encountered by a scientist who attempted to use this system may be found in the writings of the biologist (and later dissident) Zh.A. Medvedev, to which we now turn.

2.7   The Experiences of Zhores A. Medvedev

The treatment of Zhores Aleksandrovich Medvedev (Жорес Александрович Медведев) at the hands of the Soviet authorities, owing to his critique of science practice within the USSR, became something of a *cause célèbre* within international science during the late 1960s and early 1970s. Medvedev was born in Tbilisi in 1925. Following the Second World War, he forged a career as a biologist, with a particular interest in gerontology. From 1963, he worked at the Institute of Medical Radiology in Obninsk, but was dismissed from this position upon the publication in the USA of his book on Lysenkoism (Medvedev 1969). Shortly thereafter, he was arrested and detained in a psychiatric institution on account of the further publication, this time in the UK, of the texts that will be of interest to us below, although he was released following the objections of several prominent Soviet scientists (Anon 1970a, b). In 1971, Medvedev took a position at the Institute of Physiology and Biochemistry of Farm Animals in Borovsk, but soon after departed for London to take up a one-year visiting research post at the National Institute for Medical Research (Anon 1972c, 1973a). Whilst there, however, he was stripped of his Soviet

---

[85] See, respectively: Hodgson (1955), Waddington (1956), Nonweiler (1956), Seligman and Russell (1957), Deacon (1959), and Houssay (1968).

[86] See, respectively: Dyer (1962), Anon (1964), and Kato (1968).

[87] See, for example, Kline (1952, p. 83); see also the comments in Siegmund-Schultze (2014, p. 1245). Contrast this with a situation sometimes encountered in the post-Soviet world: Eastern European speakers who are fluent in Russian nevertheless insisting upon delivering their conference talks in broken English (Kryuchkova 2001, p. 413).

citizenship and thus denied re-entry to the USSR (Sweeney 1973; Anon 1973b). Despite much further protest, Medvedev was forced to remain in London, where he eventually took British nationality; his Soviet citizenship was restored in 1990 (Beeston and McEwan 1990).

In the late 1960s, Medvedev penned two essays: 'International cooperation of scientists and national frontiers' ('Международное сотрудничество ученых и национальные границы'), which describes the bureaucratic obstacles that a Soviet scientist needed to surmount in order to attend a foreign conference, and 'Secrecy of correspondence is guaranteed by law' ('Тайна переписки охраняется законом'), where Medvedev outlined his suspicions regarding the clandestine censorship of correspondence in the USSR. These essays were first circulated privately in the Soviet Union, before falling into the hands of a British publisher, which issued English translations of both (the first now under the title 'Fruitful meetings between scientists of the world') in a single volume: Medvedev (1971).[88] Although written primarily from the point of view of a biologist, these essays detail problems that were experienced by all scientists—researchers in other disciplines were, after all, subject to the same state regulation, and were users of the same postal system. I make a few comments here on Medvedev's writings, and their relevance to areas other than biology, but for a more detailed account of some of their content, the reader is referred to Hollings (2014, pp. 22–27).

As already noted, the major problems facing any Soviet scientist who wanted to travel abroad, or even to send an international letter, were bureaucratic in nature, and this is something that comes out very clearly in Medvedev's account. Indeed, what emerges from his writing is a picture of a system beset by difficulties caused not only by the highly complicated nature of Soviet bureaucracy, but also by its inherently contradictory nature: officials in different institutions had conflicting interpretations of what was or was not to be permitted. In connection with applications for foreign travel, Medvedev outlined the enormously complicated procedure that a would-be academic traveller was forced to undergo.[89] This involved the preparation of a so-called 'exit dossier', consisting of a wide range of documents, from a work history to a medical report, and also including a character reference, attesting to the applicant's "political maturity and moral stability" (Medvedev 1971, p. 13). Once compiled, the application would be passed ever upwards through various committees, ranging from discipline-specific panels to a euphemistically-titled 'exit commission' (formed of KGB officials). In principle, an application would eventually be sent for approval by the Central Committee of the Communist Party, before arriving finally at the Ministry of Foreign Affairs, who would prepare a foreign passport for the applicant and apply on their behalf to the appropriate embassy for a visa. Needless to say, only a fraction of applications would make it to this final

---

[88]Extracts from the first essay also appeared in Medvedev (1970). Medvedev's more general critique of Soviet science (Medvedev 1979) contains further details of the communications difficulties of scientists across the Iron Curtain.

[89]See Medvedev (1971, pp. 13, 195–208) or Hollings (2014, p. 23). This application procedure is also described in Levich (1976).

stage: the process could be halted at any point, with no explanation. The Western scientific literature is littered with complaints from conference organisers and delegates on the failure of Soviet invitees to appear.[90] The above procedure would often pass somewhat more smoothly if the applicant were a member of the Communist Party, and thus already deemed 'sound'. Indeed, the attendees who were dispatched from the USSR to foreign conferences were often not those scientists whom the organisers had originally invited, but delegates who had been selected by the Communist Party and/or the Academy of Sciences on the basis of their Party membership status, rather than their academic credentials—in such situations, the Soviet authorities asserted that they were in a better position than mere foreigners to judge the credentials of their own scientists and thus determine whether they were worthy of representing the USSR at international conferences.[91] Indeed, strong words were sometimes exchanged over choices of conference invitees: the invitation by a US-led organising committee of several refusenik scientists to a conference on artificial intelligence in Tbilisi in 1975 was denounced as "American provocation" (Rich 1975a, p. 5). Moreover, some of the delegates sent by the Soviet Union to foreign congresses were often not even academics, but barely-disguised KGB chaperones for the genuine scientists. This certainly did not escape the notice of delegates from other countries, who would sometimes play the game of 'spot the KGB agent'; the presence of such *faux* delegates was often noted in Western conference reports.[92]

As Medvedev related elsewhere, similar bureaucratic principles even came into play in connection with international conferences that were held within the USSR,[93] although, from around 1960, a shorter application process was adopted for travel to other socialist countries (Medvedev 1971, pp. 208–215). Thus, Czechoslovakia, for instance, became a popular venue for international conferences during the 1960s, since this was a country to which those from both East and West could travel with relative ease: witness, for example, the conferences on semiconductors (1960), plates and shells (1963), order-disorder structures (1964), and genetics (1965), as well as specialised mathematical meetings on the theories both of graphs (1963) and of semigroups (1968).[94] The situation became more problematic, however, following the Soviet invasion of 1968.[95]

---

[90] For an example from metallurgy, for instance, see Cahn (1970); for one concerning nuclear desalination, see Anon (1968b). On geophysics, see Hamblin (2000b, p. 304). For mathematical examples, see Lehto (1998, pp. 174, 189, 206). For a typical example of the general remarks that were made on this subject, see Holliday (1973). See also Krementsov (2005, pp. 70–71). A further (particularly entertaining) source in this connection is the letter Ziman (1968), on which see Hollings (2014, pp. 24–25).

[91] See, for example, Lehto (1998, Sect. 9.3).

[92] See, for example, Abelson (1971, p. 797). In his discussion of the International Conference on Peaceful Uses of Atomic Energy (Geneva, 1955), Josephson (2000, p. 174) notes the presence of "the usual KGB staffers". See also the comments in Krementsov (2007, p. 61).

[93] See Medvedev (1971, pp. 189–190); see also Anon (1972b).

[94] See, respectively: Smith (1960), Brilla and Balaš (1966), Wooster (1964), Medvedev (1971, pp. 74–80), Fiedler (1964), and Hollings (2014, Sect. 12.3).

[95] See, for example, the remarks in Anon (1968a); see also Hamblin (2000b, p. 308).

As noted above, Western conference organisers were keenly aware of, and often frustrated by, the difficulties in obtaining Soviet speakers. In the first of his essays, Medvedev recounted how, over the years, he had received exasperated letters from counterparts who were trying to bring him on visits to the West, but who had been stymied by the convoluted process of obtaining official (Soviet) permission. He recalled those instances when he had been forced to pull out of a scheduled trip at the last minute when his permission to travel had been withdrawn, with no reason given. On such occasions, Medvedev, and those placed in a similar position, would be instructed by their superiors to make up an appropriate excuse: family illness, heavy workload, etc. As examples, Medvedev reproduced some of the letters that the Soviet Academy of Medical Sciences had sent to foreign conference organisers, declining invitations on his behalf, with excuses such as

> Dr Zh. Medvedev cannot go to Vienna to take part in the Congress [of Gerontology] since at the present time he is extremely busy with a number of projects (Medvedev 1971, p. 25)

or the practically identical

> Dr Zh. Medvedev will not be able to come to England this year because of a great press of work he has to do in his laboratory.[96]

Indeed, by the time of Medvedev's writing, Soviet officialdom had been employing such excuses for many years. A number of Soviet delegates had been invited to the 1950 ICM held at Harvard, but, as we see from Table 2.1, none were able to attend. Shortly before the congress opened, the organisers received the following cablegram from the then-president of the Soviet Academy of Sciences, the physicist S.I. Vavilov:

> USSR Academy of Sciences appreciates receiving kind invitation for Soviet scientist take part in International Congress of Mathematicians to be held in Cambridge. Soviet Mathematicians being very much occupied with their regular work unable attend congress. Hope that impending congress will be significant event in mathematical science. Wish success in congress activities.[97]

Soviet mathematicians were thus at least as 'busy' as their counterparts in cell biology, for there do not appear to have been any Soviet delegates at the Seventh International Congress of Cell Biology, which was held in New Haven, Connecticut, that same year (Brooks 1950), although mathematicians and cell biologists were evidently both much busier than Soviet physiologists, since many of the latter were able to attend the Eighteenth International Physiological Congress in Copenhagen, also in 1950 (Gerard 1950). It is reasonable to suppose that the respective venues of these conferences have some significance here.

By way of concluding this section, we turn very briefly to the subject of Medvedev's second essay: postal censorship. Here, Medvedev outlined his suspicions that many of the letters he sent abroad, particularly those to the United States, were not reaching their destinations, but were instead being intercepted by the Soviet

---

[96]Medvedev (1971, p. 54). On such excuses, see also Byrnes (1976, pp. 179–180).

[97]Graves et al. (1952, vol. 1, p. 122); see also Lehto (1998, p. 89) and Kline (1952, p. 84).

authorities. His enquiries into this matter were met with the indignant assertion that postal censorship was illegal in the Soviet Union[98]—although, in fact, the examination of foreign letters, particularly scientific ones, in search of hidden messages, had been common since the mid-1930s. As for travel, bureaucracy, contradictory regulations, and the need to obtain official permissions, posed obstacles to the sending of any materials outside the USSR. The submission of papers to foreign journals was now possible, at least in principle, but the associated procedures could be difficult to negotiate. In particular, it was necessary to obtain the permission of one's university's 'First Department', a euphemistic term for the institution's KGB representative. Those assigned to assess whether papers could be sent abroad were rarely specialists, and so the securing of permissions could be particularly difficult in those disciplines with a more arcane presentation, such as mathematics. Indeed, the Soviet authorities did not merely place restrictions on material that was sent out of the USSR, but also on that coming in—more comments will be made on this issue in Chap. 3, in connection with accessibility of scientific publications.

2.8 In the Opposite Direction

The main focus of the chapter so far has been on the experiences of *Soviet* scientists, and so I have had little to say about Westerners. Indeed, I have suggested that there is more *to* say about the Soviet side, since, in many respects, Western scientists could do little but *react* to the changing policies of the USSR in connection with international communications. Nevertheless, it will be instructive to look briefly at the Western side of things, for Western scientists were by no means free of home-grown difficulties.

As we have seen, it remained broadly possible for Western scientists to travel into the USSR throughout our period of interest, although the level of difficulty in doing so varied over time. By the 1960s, however, it had almost become easy, at least in comparison to the situation in the opposite direction (Medvedev 1971, pp. 216–222). In 1961, for example, Moscow played host to delegates from 58 different countries at the Fifth International Congress of Biochemistry (Anon 1961). Moreover, as we see from Table 2.1, an enormous number of non-Soviet mathematicians attended the 1966 ICM in Moscow: over 2,000 delegates came from outside communist Central and Eastern Europe (Trostnikov 1967, p. 16). Indeed, 1966 saw the USSR host several international congresses, which similarly attracted large numbers of foreign delegates: besides mathematics, there were congresses on metallic corrosion, microbiology, food microbiology, oceanography, low-temperature physics,

---

[98] Such contradictory behaviour on the part of the Soviet authorities was subsequently seized upon by the dissident movement, which called simply for the USSR to obey its own laws. As Gessen (2011) has commented, the dissidents "demanded logic and consistency" (p. 7), so "it is perhaps no accident that the founders of the dissident movement in the Soviet Union were mathematicians and physicists" (p. 178). In this connection, see also Rich (1976).

crystallography, and psychology.[99] Several of these congresses were not without their political troubles, however (Abelson 1966). More generally, Western visitors to the USSR in the post-Stalin period typically found that their movements within the country were restricted,[100] which led in turn, in the US instance, to retaliatory restrictions on Soviet visitors to the USA (alongside restrictions imposed over genuine security concerns: see Lubrano 1981, p. 474); American attempts to negotiate a mutual lifting of these constraints came to nought (Richmond 2003, p. 26).

The submission by Westerners of papers to Soviet journals is a different matter. There do not appear to have been any particular bars to this on the Soviet side; indeed, one might speculate that Soviet editors would have welcomed Western submissions as evidence that their journals had achieved an international standing. However, such submissions appear to have been quite rare. As I will discuss in Sects. 4.2 and 4.3, they were a little more common in the 1920s and 1930s when Soviet journals accepted papers in languages other than Russian, but, with the drive towards the near-exclusive use of Russian from the late 1930s onwards, Western submissions to Soviet journals all but dried up.

To return to the issue of postal censorship, we note that, although this was certainly a bigger problem in the USSR, US scientists at one point found themselves at risk of a similar problem: there were concerns that new legislation, designed to block incoming political propaganda, would affect the receipt of Soviet scientific literature[101]—I will say a little more about this in Sect. 3.2.

In the case of the United States, other political considerations may have affected contacts with scientists on the opposite side of the Iron Curtain, although the evidence for this is nowhere near as clear-cut as one might expect. It is natural to suppose that, at the height of McCarthyism in the 1950s, contacts with Soviet scientists would have been as a red flag (no pun intended) to communist-hunting officials, particularly in light of such high-profile cases of espionage as those of Klaus Fuchs, and Julius and Ethel Rosenberg. There were certainly concerns amongst some in the USA that the purported Soviet scientists who later travelled to North America under the auspices of exchange agreements were in fact intelligence agents, seeking to steal secrets.[102] At the same time, the feeling in some quarters was that US scientists were rather too naïve in their dealings with foreign (particularly Soviet) counterparts (Schweitzer 1989, p. 153). We might thus expect to find examples of the persecution, or at least censuring, of US scientists by the authorities in connection with their Soviet contacts. However, although suggestions to this effect do indeed appear here and there in the literature, these are almost uniformly vague, and

---

[99]See, respectively: Anon (1969), Anon (1966a, b), Ingram and Roberts (1967), Charlier and Dietz (1966), Malkov (1967), Kamminga (1989, p. 599), and Rosenzweig (2000, Chap. 9).

[100]See, for example, Anon (1963b). It was observed, however, that there was not necessarily any need to travel extensively within the USSR, since most of the scientific facilities were concentrated in Moscow and Leningrad.

[101]See Byrnes (1976, pp. 122–123) and DuS (1961a, b, 1962).

[102]See, for example, Schweitzer (1989, Chap. 7). Indeed, these concerns may not have been entirely unfounded: see the brief comment on p. 29.

rarely give details on the precise nature of the imputed harassment.[103] The authoritative source on McCarthyism and US academia, namely Ellen Schrecker's *No Ivory Tower* (Schrecker 1986), says little on this matter: the persecuted scientists profiled by Schrecker were all singled out for their left-leaning politics—although it must be observed that the records available to Schrecker were in some cases quite patchy (Gruber 1987). One can easily imagine that any efforts made by such scientists to contact Soviet counterparts would not have helped their cause,[104] but the question of whether there was any persecution of scientists of 'sound' politics who attempted to contact Soviet colleagues remains open. The US (or, more generally, Western) authorities may indeed have harboured suspicions of scientists with contacts behind the Iron Curtain, but examples of their having acted on such doubts are rare.[105] It is possible that those scientists who have claimed to have had difficulties with the authorities when attempting to contact Soviet counterparts may have been exaggerating their experiences in order to present themselves in a slightly subversive light, and thus distance themselves from the excesses of the Cold War political climate: as noted above, their comments are rarely specific, and appear to be more in the nature of innuendo. Indeed, such comments clash with those made elsewhere. Take, for instance, the following general remarks made in connection with the Danish marine biologist Anton Bruun, whose contacts with both American and Soviet counterparts made him a conduit for the communication of oceanographic research:

> For the West, Soviet science became a source of fascination and fear, its shadowy nature only encouraging curiosity. Individuals who could chart its contours with greater clarity thus possessed a valuable currency.[106]

The peripatetic Hungarian mathematician Paul Erdős served as a similar point of contact for mathematics, although he did encounter problems when the United

---

[103]See, for example, the comment that officials at the US State Department "often regarded the efforts of scientists to maintain international contacts as synonymous with communist sympathies" in Doel and Needell (1997, p. 69); see also Doel et al. (2005, p. 67). For such suspicions within the context of the Manhattan Project, see Schrecker (1986, p. 133ff). Moving away from the United States, we have the obscurely-referenced "retrospectively amusing difficulties with the authorities" apparently experienced by the British mathematician G.B. Preston in his efforts to establish contacts with Eastern-bloc colleagues (Howie 1995, p. 269). Niels Bohr was at one point considered a security risk because of his contacts with Soviet physicists (Nielsen and Knudsen 2013, p. 322); see also Aaserud (1999, pp. 32–33) and Knudsen and Nielsen (2012).

[104]One shudders to contemplate, for example, the impression created by the failure to appear before the House Un-American Activities Committee of the US mathematician Stephen Smale by reason of his attendance of the 1966 Moscow ICM; see Greenberg (1966) and Smale (1984).

[105]Examples are provided by the FBI surveillance of certain scientists with Soviet or communist-bloc contacts: see Krementsov (2002, p. 109) and Kerber (2012, p. 234). These included US-based German rocket scientists with connections in the newly-created East Germany (Cadbury 2005, pp. 132–133). Even if the US authorities rarely acted against academics with foreign contacts, this did not stop some scientists fearing the backlash that communications even with other Western nations might bring: for example, Ellsworth Dougherty, a biophysicist with interests in atomic science, refused to share work with a British colleague in order to avoid any appearance of creating a security breach; see Manzione (2000, p. 40).

[106]Roberts (2013, p. 251); see also Heymann and Martin-Nielsen (2013, p. 232).

States refused him re-entry following his attendance of the 1954 Amsterdam ICM (Hoffman 1998, pp. 128–129). One of Erdős' biographers suggests that Erdős' contacts with the Chinese mathematician Lo Ken Hua may also have given the US authorities 'reason' to regard him with suspicion (Schechter 1998, pp. 162–167).

Erdős was not the only foreign traveller to experience difficulties regarding US visas.[107] The early 1950s saw greater restrictions imposed on visitors to the United States (termed a 'paper curtain' by one author: Anon 1955), the result of which, coupled with the inability of some US scientists to obtain passports (see below), was effectively to isolate American science, much to the concern of many US scientists, who feared that such a segregation from world science would cost the United States its ascendancy (Doel et al. 2005, p. 68). The difficulties of getting foreign (particularly, Soviet-bloc) speakers to conferences in the USA meant that very few international scientific congresses were held there during the late 1940s and early 1950s (Manzione 2000, p. 39). A notable exception is the Harvard ICM of 1950. However, not only were there no Soviet delegates present at this congress (see Table 2.1), there were also no delegates from anywhere else within the Eastern bloc—though the congress proceedings firmly absolve the US government of any culpability in this regard:

> In attempting to maintain the non-political nature of the Congress, many serious difficulties had to be overcome. In the solution of these problems, officers of the Congress found the various officials of the Department of State most sympathetic and helpful. As a part of the effort to keep the Congress apolitical, they tried to secure a visa for every mathematician who notified them about any visa difficulties before cancelling his passage. As far as they know only one mathematician from any independent nation was prevented from attending the Congress because he failed to pass a political test and this man did not notify the officers of the Congress about his difficulties. Only two mathematicians from occupied countries failed to secure visas. Mathematicians from behind the Iron Curtain were uniformly prevented from attending the Congress by their own governments which generally refused to issue passports to them for the trip to the Congress. Their non-attendance was not due to any action of the United States Government. (Graves et al. 1952, vol. 1, p. 122)

Even if they played no part in the difficulties of the 1950 ICM, wider US actions in connection with visas were subject to international condemnation and also provoked many an angry reaction from American scientists,[108] leading, for example, to the formation by the Federation of American Scientists of a Committee on Visa Problems, whose remit was to investigate the difficulties experienced by would-be foreign visitors and to lobby the US government for change (Doel et al. 2005, p. 68). Little appears to have been achieved, however: by 1955, four international scientific meetings that had been scheduled to take place within the USA had been moved elsewhere (Doel et al. 2005, p. 68). Moreover, US scientists were forced to turn down a proposed exchange programme with the USSR (Manzione 2000, p. 41). Only gradually did the problems ease—but they didn't disappear altogether: as late

---

[107]Paul Dirac was another prominent scientist who was denied a visa to visit the USA, probably because of his several pre-war visits to the USSR; see Dalitz and Peierls (1986, p. 158).

[108]See, for example, Manzione (2000, p. 40). On the USA's visa restrictions, see Bok (1955), and also the articles in vol. 8, no. 7 (October 1952) of the *Bulletin of the Atomic Scientists*.

as 1980, visas were still being refused to certain Soviet scientists (Lubrano 1981, p. 474). Nevertheless, conference organisers were often able to carry out careful negotiations with the US State Department. One such case of mediation was that initiated by the US National Committee of the IAU (p. 31), which ensured the presence at the IAU's Berkeley General Assembly of 1961 of many foreign delegates who might not otherwise have been able to attend (Blaauw 1994, Sect. 10.d). It should be noted, however, that not all US scientists had condemned their government's actions: some saw the restrictions on travel as "a necessary sacrifice to win the cold war" (Doel et al. 2005, p. 72).

I conclude this section with some reflections on the place of McCarthyism within the present discussion. It has been suggested that the effects of communist witchhunts on the American scientific community were not as pronounced as is commonly supposed: that there were a few high-profile instances of persecution, particularly where the scientist in question had engaged in defence work (such as was the case with J. Robert Oppenheimer—see, for example, Wolfe 2013, pp. 21, 36–37), but that the overall impact was slight (Schrecker 1998, pp. 404–407). Nevertheless, leaving issues of international communication aside for the moment, and also stepping beyond the purely scientific realm to consider academia more generally, we may draw loose parallels between the treatment of some US academics under McCarthyism, and events in the USSR such as the 'Luzin affair' (Sect. 2.3). In both instances, the state signalled its disapproval of contacts between its scholars and those on the opposite side of the Iron Curtain. As we have noted, travel restrictions were imposed both into and out of the USA. Those to which US academics were subjected were of course milder than those under which Soviet researchers laboured, but they were by no means trivial—many US scholars whose political allegiances came under question had great difficulties in obtaining foreign passports[109]; in one instance, perhaps in a faint echo of the USSR's predilection for sending only 'politically sound' academics abroad, the US State Department even offered to send a different American scientist for a job in India after it had denied a passport to the chosen candidate (Schrecker 1986, p. 297). Pessimistic about the probable outcome, many 'tainted' US academics simply did not attempt to travel (Schrecker 1986, p. 296). In the opposite direction, the bureaucracy surrounding US visa restrictions on foreign scientists appears in some cases to have mirrored that touched upon in Sect. 2.7 (Nassau 1956; Anon 1955). The ugly shade of career ambition also played a role in both the USA and the USSR: just as Luzin's disgruntled students appear to have sought their own advancement through his downfall, so too did some US academics view cooperation with the House Un-American Activities Committee as an easy path to career development (Schrecker 1986, p. 195). Finally, those who had been maltreated by McCarthy (and others) eventually underwent a process of restitution that was reminiscent of the post-Stalinist Soviet practice of 'rehabilitation', whereby those persons (still alive or not) who had previously been persecuted were restored to a state of political acceptance (Schrecker 1986, p. 338). Overall,

---

[109]See the various instances cited by Schrecker (1986, pp. 145, 147, 168, 197, 278, 296–297).

the scope of the victimisation of US nationals during the relevant period hardly compares with that of Soviet citizens, but it is nevertheless interesting to note the elements that the two situations had in common.[110]

2.9 Concluding Remarks on Personal Communications

In spite of many continuing difficulties, there was, by the 1960s, a regular exchange of scientific knowledge across the Iron Curtain, in both directions; the *ad hoc* exchange programmes of the preceding decades were also being replaced by larger-scale, centrally-organised schemes, first on well-defined, quite rigidly-specified projects, and then programmes of a more flexible nature.[111] The situation in this period is summarised by Claude Debru in the following terms:

> scientists from the Soviet union and satellite countries were able to communicate with their colleagues from the Western world even in the 1950s and 1960s in spite of the mental walls erected by communist authorities in the Eastern block [*sic*] countries, and in spite of occasional difficulties. The situation of individual scientists did, however, vary depending on local circumstances, on the various disciplines and on the big institutions. (Debru 2013, pp. 64–65)

Scientists from one side of the Iron Curtain were travelling quite regularly to attend conferences on the other; witness, for example, the significant US presence at the Fifteenth General Assembly of the International Union of Geodesy and Geophysics in Moscow in August 1971, at which it was noted that the Soviet delegates interacted more freely with foreigners than they had at an earlier meeting in Helsinki in 1960 (Abelson 1971, p. 797). In general, conference delegates travelled not merely for scientific reasons, but were also motivated simply by curiosity. This certainly seems to have been the case for many Western delegates at the 1966 Moscow ICM (Lehto 1998, Sects. 8.1–8.2). In order to satisfy the further curiosity of their non-travelling colleagues, returning conference attendees would often pen reports on what they had seen—such reports will be of particular interest to us in Sect. 3.5.

The situation, however, was not utopian, for international communications were still plagued, on occasion, by the same difficulties that have been described here.[112] For example, of the 36 Soviet mathematicians invited to attend the 1986 ICM in Berkeley, California, only 19 were, ultimately, able to attend.[113] Moreover, the US State Department continued occasionally to hinder international exchanges by placing severe travel restrictions on Soviet visitors to the USA (Shapley 1974). Canadian

---

[110]See also the similar comments in Gordin et al. (2003, pp. 50–53).

[111]See the various sources cited in the first paragraph of Chap.1, which deal also with the new exchange programmes that continued to be negotiated right up until the end of the Soviet era; on these, see also Korneyev (1977, pp. 320–326). On post-Soviet US-Russian scientific exchanges, see Schweitzer (1997).

[112]See, for example, Medvedev (1979, pp. 152–153) or Reid (1977).

[113]See Nathanson (1986). As we see from Table 2.1, the congress was attended by a further 38 Soviet delegates who had not been specifically invited.

visas were denied to Soviet delegates hoping to attend the 1984 General Assembly of the ICSU in Ottawa (Greenaway 1996, p. 102). On the other side of the Iron Curtain, visiting US scientists complained of harassment at the hands of the KGB.[114] Indeed, reports such as these, coupled with concerns about the conditions of life in the USSR, seem to have discouraged some US scientists from participating in the various exchange programmes that were emerging[115]: the impressions recorded by those scientists who did visit the USSR could be very mixed (Schweitzer 1989, pp. 183–185). As noted in Sect. 2.6, Western unease about human rights issues in the USSR provided a stumbling block when it came to the smooth operation of scientific exchanges.[116] Several cultural exchanges were curtailed, for example, following the Soviet invasion of Afghanistan.[117]

Nevertheless, in spite of continuing problems, scientific contacts between East and West were, by the 1970s and 1980s, better than they had ever been. Westerners in particular continued to strive for improved connections with, and broader understanding of, the Soviet Union, in connection with science and technology, and also more generally (see also the comments in Sect. 3.5). In a US text of 1976, for example, we find reference to

> the relentless American interest in increasing knowledge and understanding of Russia and Eastern Europe, which remains even now quite deficient. (Byrnes 1976, p. 3)

The same source contains a lamentation of "[t]he absence of a powerful parallel Soviet interest in increasing learning and insight concerning the United States", which we should probably interpret as referring to the resistance of Soviet officials to the import of general American culture. We have seen that, following Stalin's death, the USSR engaged enthusiastically in scientific and technological exchanges with the West, in order to gain access to Western innovations, although the Soviet state remained wary of wider Western influences reaching the people.

Closer East-West ties effected a curious change in official Soviet attitudes: in contrast to the earlier condemnation of 'servility to the West' and the attitudes surrounding the 'Luzin affair', it was now observed by a US commentator that

> Soviet scientists take great pride in their publications in Western journals, publications which are often facilitated through collaborative efforts with Western colleagues. (Schweitzer 1989, p. 181)

Indeed, honorary membership of foreign learned societies even became a positive factor in securing promotions within Soviet academia (Schweitzer 1989, p. 182).

---

[114]See Byrnes (1976, pp. 192–198). Indeed, even from the early years of the USSR, VOKS had aided the Soviet secret police in keeping track of foreign visitors; see David-Fox (2012, pp. 58–59). Conversely, Soviet visitors to the USA had sometimes been kept under surveillance by the FBI; see Richmond (2003, p. 28).

[115]See Byrnes (1976, p. 115); for reports (mostly) of a more positive nature, see Kuznick (1987, pp. 112–143).

[116]See the references in note 75 on p. 30.

[117]See, for example, Lubrano (1981), Katz et al. (1980, p. 6), and Anon (1984).

Recall from Sect. 2.6 the remark of a Western observer that Soviet scientists were, collectively, 'more open' to entertaining foreign ideas than perhaps were other segments of the Soviet population. Indeed, this open-mindedness on the part of Soviet scientists, and academics more generally, fuelled the hopes of Westerners that increasing contacts with the USSR, both cultural and academic, might gradually serve to influence Soviet policy and attitudes. Scientific contacts played a particularly prominent role in this connection, as Graham (1998, pp. 32–33) has remarked:

> [s]cience and technology have acted powerfully as moderating influences, as forces pulling Russia towards the West, as factors reducing the differences between Russia and the West.

Some authors[118] have argued that the gradual 'Westernisation' of Soviet academics through visits to the USA, for example, may have contributed first to the opening up of the USSR during the 1980s, and then to its eventual collapse:

> Among the thousands of Soviet and East European academics and intellectuals who were exchange participants in the United States and Western Europe during … the Cold War, many became members of what, in retrospect, turned out to be underground establishments. They were well-placed individuals, members of the political and academic elites, who began as loyalists but whose outside experiences sensitized them to the need for basic change. Together with the more radical political and cultural dissidents, towards whom they were ambivalent or hostile, they turned out to be agents of change who played a key part, sometimes unintentional, in the demise of European Communism. (Kassof 1995, p. 263)

The author Yale Richmond (2003, pp. 22–47) has recorded several examples of Russians who visited the USA as students and who, in later years, became prominent Soviet policy-makers. Although, in most cases, these people were not turned against the USSR, they had nevertheless enjoyed a glimpse of a different system, and went on to introduce a more liberal element into Soviet politics. Official Soviet fears of 'cultural pollution' thus appear to have been justified.

References

Aaserud, F.: The scientist and the statesmen: Niels Bohr's political crusade during World War II. Hist. Stud. Phys. Biol. Sci. **30**(1), 1–47 (1999)

Abelson, P.H.: International meetings. Science 154(3747), 21 Oct, 341 (1966)

Abelson, P.H.: Geophysicists in Moscow: signs of easier relations. Science 173(3999), 27 Aug, 797–800 (1971)

Aleksandrov, D.A.: Why Soviet scientists stopped published abroad: the establishment of the self-sufficiency and isolation of Soviet science 1914–1940. Voprosy istor. estest. tekhn. **3**, 4–24 (1996) (in Russian)

Aleksandrov, P.S.: First International Topological Congress in Moscow. Uspekhi mat. nauk, no. 1, 260–262 (1936) (in Russian)

Aleksandrov, P.S.: Pages from an autobiography. Uspekhi mat. nauk 34(6), 219–249 (1979); ibid. 35(3), 241–278 (1979) (both in Russian); English trans.: Russian Math. Surveys 34(6), 267–302 (1979); ibid. 35(3), 315–358 (1979)

---

[118]For instance, Kassof (1995), Graham (1998, Chap. 2), and Richmond (2003, Chaps. 4 and 5).

Anon: The International Geological Congress. Nature **38**(986), 20 Sept, 503–506 (1888)

Anon: Scientific news. Amer. Naturalist 27(320), Aug, 764–767 (1893)

Anon: Physical science at the British Association. Nature 108(2718), 1 Dec, 448–450 (1921)

Anon: The Russian Academy of Sciences. Nature 116(2916), 19 Sept, 448–449 (1925)

Anon: Soviet mathematicians, support your journal! Mat. sb. 38(3–4), 1 (1931) (in Russian)

Anon: Notes. Bull. Amer. Math. Soc. 40(9), 648–651 (1934)

Anon: Proceedings of the Second All-Union Mathematical Congress, Leningrad, 24–30 June 1934. Izdat. Akad. nauk SSSR, Moscow/Leningrad (1935a) (in Russian)

Anon: International Physiological Congress in Moscow and Leningrad. Lancet 226(5843), 24 Aug, 446–448 (1935b); ibid. 226(5845), 7 Sept, 575–576 (1935b)

Anon: Proceedings of the First All-Union Congress of Mathematicians (Kharkov, 1930). Glav. Red. Obshchetekhn. Lit. Nomogr., Moscow/Leningrad (1936) (in Russian)

Anon: The Warsaw Conference on Modern Physics. Science 88(2273), 22 Jul 1938, 76

Anon: American mathematicians and the U.S.S.R. Nature 148(3758), 8 Nov, 560 (1941a); also appeared in print as: The mathematicians of America and of Soviet Russia. Science 94(2441), 10 Oct, 340 (1941a)

Anon: American mathematicians and the U.S.S.R. Nature 148(3761), 29 Nov, 657 (1941b); also appeared in print as: Greetings of Soviet mathematicians to American mathematicians. Science 94(2444), 31 Oct, 409 (1941b)

Anon: Great Britain and the U.S.S.R. Nature 148(3744), 2 Aug, 135–136 (1941c); ibid., 9 Aug, 160 (1941)

Anon: War communique [sic] from Leningrad scientists. Soviet War News, no. 28, 12 Aug, 3 (1941d)

Anon: New discoveries by Soviet astronomers. Soviet War News, no. 28, 12 Aug, 3 (1941e)

Anon: Soviet scientists — a part of the armed forces. Soviet War News, no. 42, 28 Aug, 2 (1941f)

Anon: Anglo-Soviet Medical Committee. Soviet War News, no. 49, 5 Sept, 4 (1941g)

Anon: Professors with hand grenades. Soviet War News, no. 53, 10 Sept, 3 (1941h)

Anon: What Leningrad scientists are doing. Soviet War News, no. 80, 12 Oct, 4 (1941i)

Anon: Anglo-Russian alliance in medicine. Lancet 238(6160), 20 Sept, 344 (1941j)

Anon: Anglo-Soviet Medical Committee. Brit. Med. J. 2(4216), 25 Oct, 590 (1941k)

Anon: Scientific co-operation between Great Britain and the U.S.S.R. Nature 149(3776), 14 Mar, 297 (1942a)

Anon: Race and fascism. Nature 149(3781), 18 Apr, 426–427 (1942b)

Anon: Science and technology in the Soviet Union. Nature 149(3785), 16 May, 545–547 (1942c)

Anon: The U.S.S.R. Academy of Science and the Royal Society. Nature 149(3789), 13 Jun, 663 (1942d)

Anon: Exchange of scientific information with the U.S.S.R. Nature 149(3789), 13 Jun, 663 (1942e)

Anon: British and Russian naturalists. Nature 150(3795), 25 Jul, 117 (1942f)

Anon: Anglo-Soviet Scientific Collaboration Committee. Nature 150(3801), 5 Sept, 285–286 (1942g)

Anon: Science and Technology in the Soviet Union. Papers Read at the Symposium at Easter, 1942, held under the Auspices of The Faculty of Science of Marx House. Science Services Ltd. (1942h)

Anon: Russian interchange. Lancet 239(6183), 28 Feb, 263 (1942i)

Anon: A book for Russian colleagues. Lancet 240(6223), 5 Dec, 674 (1942j)

Anon: Message received by the American Association of Scientific Workers from the Soviet Scientists [sic] Antifascist Committee. Science 97(2511), 12 Feb, 162–163 (1943a)

Anon: The work of Soviet astronomers at Leningrad during the siege. Science 97(2515), 12 Mar, 237 (1943b)

Anon: British and Russian men of science: exchange of greetings. Nature 152(3846), 17 Jul, 70 (1943c)

Anon: Science in Soviet Russia: Papers presented at Congress [sic] of American-Soviet Friendship, New York City, November 7, 1943, under the auspices of the National Council of American-Soviet Friendship. Jaques Cattell Press, Lancaster, PA (1944)

Anon: Anniversary of the Academy of Sciences of the U.S.S.R. Science 101(2633), 15 Jun, 603 (1945a)

Anon: Anniversary of the Academy of Sciences of the Soviet Union. Science 102(2638), 20 Jul, 58 (1945b)

Anon: Science in Peace. Nature 155(3931), 3 Mar, 260–262 (1945c)

Anon: British representatives at Soviet Academy celebrations. Nature 155(3946), 16 Jun, 721 (1945d)

Anon: Solar energy for Soviet economy: first 'solar kitchens' and 'solar laundries'. Soviet War News, no. 1114, 20 Mar, 4 (1945e)

Anon: 220th anniversary of Academy of Sciences of the U.S.S.R.: jubilee session in June. Soviet War News, no. 1163, 24 May, 1 (1945f)

Anon: Jubilee of Soviet Academy of Sciences: 44 British scientists invited. Soviet War News, no. 1166, 28 May, 2 (1945g)

Anon: British scientists arrive in Moscow. Soviet War News, no. 1183, 16 Jun, 1 (1945h)

Anon: Scientists in session. Soviet War News, no. 1185, 19 Jun, 1 (1945i)

Anon: Medical exchange with Russia ends. NY Times, 19 Nov, 19 (1948)

Anon: Iron Curtain breaks. Science Newsl. 65(22), 29 May, 351 (1954)

Anon: Visa problems of scientists. Science Newsl. 68(25), 17 Dec, 386 (1955)

Anon: Biochemistry in Russia: impressions gained at the Fifth International Congress of Biochemistry in Moscow. Brit. Med. J. 2(5253), 9 Sept, 701–703 (1961)

Anon: US-USSR scholarly exchange program. ACLS Newsl. 13(4), 1–4 (1962a)

Anon: The largest cancer congress. Brit. Med. J. 2(5301), 11 Aug, 405–406 (1962b)

Anon: ACLS exchange program with the Academy of Sciences of the USSR. ACLS Newsl. 14(2), 15 (1963a)

Anon: Program of exchanges of scholars between the U.S. and the U.S.S.R. ACLS Newsl. 14(7), 8–11 (1963b)

Anon: A British view of Soviet research administration. New Scientist, no. 368, 5 Dec, 586–587 (1963c)

Anon: Sixth International Congress of Biochemistry, July 16 to August 1, 1964, New York City. Proceedings of the Plenary Sessions and the Program. International Union of Biochemistry (1964)

Anon: Microbiologists in Moscow. Nature 211(5052), 27 Aug, 900–901 (1966a)

Anon: International Association of Microbiological Societies, 9th International Congress for Microbiology, Moscow, 1966. Pergamon Press, Oxford (1966b)

Anon: Czechoslovak conference: biology of bats. Nature 220(5163), 12 Oct, 117 (1968a)

Anon: Water: nuclear desalination. Nature 220(5174), 28 Dec, 1274 (1968b)

Anon: Proceedings of the Third International Congress on Metallic Corrosion, Moscow, 1966. Mir, Moscow (1969)

Anon: Sanity restored. Nature 226(5252), 27 Jun, 1188 (1970a)

Anon: Clear voice from the East. Nature 227(5264), 19 Sept, 1177–1178 (1970b)

Anon: Scientific accord in Moscow. Nature 237(5353), 2 Jun, 247–248 (1972a)

Anon: Zhores Medvedev and the reputation of Russian science. Nature 238(5359), 14 Jul, 61–62 (1972b)

Anon: Medvedev given two year passport. Times (London), 27 Nov, 14 (1972c)

Anon: Dr Medvedev in London. Nature 241(5386), 19 Jan, 154 (1973a)

Anon: Sad case of Dr Zhores Medvedev. Nature 244(5416), 17 Aug, 379 (1973b)

Anon: Scientific exchanges must serve science, not the Soviet government. Nature 283(5746), 31 Jan, 415 (1980)

Anon: How to order an East-West thaw. Nature 309(5971), 28 Jun, 735–736 (1984)

Avery, D.: Allied scientific co-operation and Soviet espionage in Canada, 1941–45. Intell. Nat. Security **8**(3), 100–128 (1993)

Aver'yanov, A.G., Korotkevich, Ye.S.: Glaciation of Antarctica (study results and prospects). Polar Geogr. **2**(3), 154–163 (1978)

Babes, M., Kaeser, M.-A. (eds.): Archaeologists Without Boundaries: Towards a History of International Archaeological Congresses (1866–2006). Archaeopress, Oxford (2009)

Baskerville, C.: International congresses. Science **32**(828), 11 Nov, 652–659 (1910)

Bateson, W.: Science in Russia. Nature **116**(2923), 7 Nov, 681–683 (1925)

Beardsley, E.H.: Secrets between friends: applied science exchange between the Western Allies and the Soviet Union during World War II. Social Stud. Sci. **7**, 447–473 (1977)

Beeston, N., McEwan, A.: Soviet exile gets back his citizenship. Times (London), 2 Jul, 9 (1990)

Bernal, J.D.: Scientists at the front. Brit. soyuzn., no. 2(74), 9 Jan, 6 (1944) (in Russian)

Blaauw, A.: History of the IAU: The Birth and First Half-Century of the International Astronomical Union. Springer (1994)

Bok, B.J.: Science in international cooperation. Science **121**(3155), 17 Jun, 843–847 (1955)

Brilla, J., Balaš, J. (eds.): Theory of Plates and Shells: Selected Papers Presented to the Conference on the Theory of Two- and Three-dimensional Structures held at Smolenice (Slovakia), October 1st–5th, 1963. Slovak Acad. Sci., Bratislava (1966)

Brooks, J.L.: VIIth International Congress of the International Society for Cell Biology. Science **112**(2921), 22 Dec, 769–770 (1950)

Bu, L.: Educational exchange and cultural diplomacy in the Cold War. J. Amer. Studies **33**(3), 393–415 (1999)

Bunbury, D.E.: The formation of the Anglo-Soviet Medical Committee. Postgrad. Med. J. **18**(194), 6–7 (1942)

Burns, D.T., Deelstra, H.: The origins and impact of the International Congresses of Applied Chemistry, 1894–1912. Microchim. Acta **172**, 277–283 (2011)

Byrnes, R.F.: Soviet-American Academic Exchanges, 1958–1975. Indiana Univ. Press (1976)

Cadbury, D.: Space Race: The Untold Story of Two Rivals and Their Struggle for the Moon. Fourth Estate, London (2005)

Cahn, R.W.: Restrictions on Soviet scientists. Nature **228**(5270), 31 Oct, 485 (1970)

Cannon, W.B.: Foreword. Amer. Rev. Soviet Med. **1**(1), 5–6 (1943)

Carling, E.R.: Medical and surgical achievement in the U.S.S.R. during the war. Nature **153**(3884), 8 Apr, 419–424 (1944)

Carlson, E.A.: Genes, Radiation and Society: The Life and Work of H. J. Muller. Cornell Univ. Press (1981)

Carlson, E.A.: Speaking out about the social implications of science: the uneven legacy of H. J. Muller. Genetics **187**(1), 1–7 (2011)

Case, C.C., Dunbar, C.O., Howe, H.V., Howell, B.F.: International Paleontological Union. J. Paleontology **12**(3), 303–304 (1938)

Chamberlin, W.H.: Foreword. Russian Rev. **1**(1), 1–5 (1941)

Charlier, R.H., Dietz, R.S.: Oceanography: two reports on the recent International Congress in Moscow. Science **153**(3742), 16 Sept, 1421–1428 (1966)

Cock, A.G.: Chauvinism and internationalism in science: the International Research Council, 1919–1926. Notes Records Roy. Soc. London **37**(2), 249–288 (1983)

Coleman, A.P.: A Report on the Status of Russian and Other Slavic and East European Languages in the Educational Institutions of the United States, its Territories, Possessions and Mandates, with Additional Data on Similar Studies in Canada and Latin America. American Association of Teachers of Slavic and East European Languages, Columbia Univ., New York (1948)

Congrès: XIme Congrès international de navigation, Saint-Pétersbourg 1908. Compte rendu des travaux du congrès. Imprimerie des travaux publics, Bruxelles (1908)

Congrès: XI International Navigation Congress, Saint Petersburg, 1908. Proceedings of the congress. Lectures and communications by foreign members of the congress on issues relating to maritime navigation. Saint Petersburg (1910) (in Russian)

Crawford, E.: Internationalism in science as a casualty of the First World War: relations between German and Allied scientists as reected in nominations for the Nobel Prizes in physics and chemistry. Social Sci. Inform. **27**(2), 163–201 (1988)

Crawford, E.: Nationalism and Internationalism in Science, 1880–1939: Four Studies of the Nobel Population. Cambridge Univ. Press (1992)

Crew, F.A.E.: Seventh International Genetical Congress. Nature 144(3646), 16 Sept, 496–498 (1939)

Dale, H.: Isaac Newton, 1642–1727. Notes Records Roy. Soc. London **4**(2), 146–161 (1946)

Dalitz, R.H., Peierls, R.: Paul Adrien Maurice Dirac. 8 August 1902–20 October 1984. Biogr. Mem. Fellows Roy. Soc. 32, 138–185 (1986)

Dalmedico, A.D.: Mathematics in the twentieth century. In: Krige and Pestre (1997), pp. 651–667

David-Fox, M.: From illusory 'society' to intellectual 'public': VOKS, international travel and party-intelligentsia relations in the interwar period. Contemp. Europ. Hist. **11**(1), 7–32 (2002)

David-Fox, M.: Showcasing the Great Experiment: Cultural Diplomacy and Western Visitors to the Soviet Union, 1921–1941. Oxford Univ. Press (2012)

Dawson of Penn, Gye, W.E., Hopkins, F.G., Manson-Bahr, P., Ryle, J.A., Shaw, W.F., Webb-Johnson, A., Wilson, C.M.: Medical coöperation with Russia. Lancet 238(6160), 20 Sept, 352 (1941a)

Dawson of Penn, Gye, W.E., Hopkins, F.G., Manson-Bahr, P., Ryle, J.A., Shaw, W.F., Webb-Johnson, A., Wilson, C.M.: Anglo-Soviet Medical Committee. Brit. Med. J. 2(4211), 20 Sept, 424 (1941b)

Deacon, G.E.R.: International Oceanographic Conference. Nature 184(4699), 21 Nov, 1605–1606 (1959)

de Andrada, E.N.: Science serves at the front. Brit. soyuzn., no. 11(83), 12 Mar, 7 (1944) (in Russian)

Debru, C.: Postwar science in divided Europe: a continuing cooperation. Centaurus **55**(1), 62–69 (2013)

Deemer, R.B., Dawson, P.R., Merz, A.R. (eds.): Proceedings and Papers of the First International Congress of Soil Science, June 13–22, 1927, Washington. American Organizing Committee of the First International Congress of Soil Science (1928)

Demidov, S.S., Esakov, V.D.: The 'case of Academician N. N. Luzin' in light of the Stalinist reform of Soviet science. Istor.-mat. issled., no. 4(39), 156–171 (1999) (in Russian)

Demidov, S.S., Levshin, B.V. (eds.): The Case of Academician Nikolai Nikolaevich Luzin. Russian Christian Humanitarian Institute, Saint Petersburg (1999) (in Russian)

Demidov, S.S., Tokareva, T.A.: The formation of the Soviet mathematical school. Istor.-mat. issled., no. 10(45), 142–159 (2005) (in Russian)

de Milt, C.: The congress at Karlsruhe. J. Chem. Ed. **28**(8), 421–425 (1951)

Denny, H.: Russian astronomer is accused of 'servility' to foreign science. NY Times, 20 Jul, 1 (1936)

Doel, R.E., Hoffmann, D., Krementsov, K.: National states and international Science: a comparative history of international science congresses in Hitler's Germany, Stalin's Russia, and Cold-War United States. Osiris **20**, 49–76 (2005)

Doel, R.E., Needell, A.A.: Science, scientists, and the CIA: balancing international ideals, national needs, and professional opportunities. Intell. Nat. Security **12**(1), 59–81 (1997)

Dunn, L.C.: Soviet biology. Science 99(2561), 28 Jan, 65–67 (1944)

DuS., G.: 'Neither Snow Nor Rain Nor . . .'. Science 133(3452), 24 Feb, 549 (1961a)

DuS., G.: The reluctant dragon. Science 133(3465), 26 May, 1677 (1961b)

DuS., G.: Postal censorship. Science **135**(3507), 16 March, 877 (1962)

Dyer, E.R., Jr.: International co-operation in space research (I). Nature 195(4840), 4 Aug, 420–423 (1962); (II). ibid. 195(4841), 11 Aug, 532–537 (1962)

Edelson, B.I., Townsend, A.: U.S.-Soviet cooperation: opportunities in space. SAIS Rev. **9**(1), 183–197 (1989)

Fedorov, K.A.: Soviet-French scientifico-technical connections. Voprosy istor. estest. tekhn., no. 1, 124–130 (1984) (in Russian)

Fenn, W.O., Franklin, K.J., Zotterman, Y. (eds.): History of the International Congresses of Physiological Sciences 1889–1968. Amer. Physiol. Soc., Baltimore, MD (1968)

Fiedler, M. (ed.): Theory of Graphs and its Application. Proceedings of the Symposium held in Smolenice in June 1963. Czechoslovak Acad. Sci., Prague (1964)

Foreign Office (UK): Report of the Interdepartmental Commission of Enquiry on Oriental, Slavonic East European and African Studies. HMSO, London (1947)

Forman, P.: Scientific internationalism and the Weimar physicists: the ideology and its manipulation in Germany after World War I. Isis **64**(2), 150–180 (1973)

Francis, W.L.: The handling of scientific and technical information in the USSR: a report on the DSIR-Aslib visit. Aslib Proc. **15**(12), 364–373 (1963)

Franklin, K.J.: A short history of the International Congresses of Physiologists. Ann. Sci. **3**(3), 241–335 (1938); reproduced in Fenn et al. (1968)

Frumkin, A.N.: Chemistry versus fascism. Soviet War News, no. 89, 22 Aug, 3 (1941)

Frye, R.N.: Review: Report of the Interdepartmental Commission of Enquiry on Oriental, Slavonic, East European and African Studies. J. Amer. Oriental Soc. **67**(4), 333–334 (1947)

Furaev, V.K.: Soviet-American scientific and cultural relations (1924–1933). Voprosy istorii, no. 3, 41–57 (1974) (in Russian); English trans.: Soviet Stud. Hist. 14(3), 46–75 (1975–1976)

G.B.: International Conference on Topology. Science 82(2134), 22 Nov, 483 (1935)

Gerard, R.W.: The Eighteenth International Physiological Congress. Science 112(2921), 22 Dec, 767 (1950)

Gerovitch, S.: From Newpeak to Cyberspeak: A History of Soviet Cybernetics. MIT Press (2002)

Gessen, M.: Perfect Rigour: A Genius and the Mathematical Breakthrough of the Century. Icon Books (2011)

Gordin, M.D.: The Heidelberg Circle: German inections on the professionalization of Russian chemistry in the 1860s. Osiris **23**(1), 23–49 (2008)

Gordin, M.D.: The Soviet science system. The Point, no. 8, 118–127 (2014)

Gordin, M.D.: Scientific Babel: The Language of Science from the Fall of Latin to the Rise of English. Profile Books (2015)

Gordin, M., Grunden, W., Walker, M., Wang, Z.: "Ideologically correct" science. In: Walker, M. (ed.) Science and Ideology: A Comparative History, pp. 38–65. Routledge, London and New York (2003)

Gordon, W.T.: The Seventeenth International Geological Congress. Nature 140(3549), 6 Nov, 789–791 (1937)

Graham, L.R.: Science and Philosophy in the Soviet Union. Alfred A. Knopf, New York (1972)

Graham, L.R.: Science in Russia and the Soviet Union: A Short History. Cambridge Univ. Press (1993)

Graham, L.R.: What Have We Learned About Science and Technology from the Russian Experience? Stanford Univ. Press (1998)

Graham, L., Kantor, J.-M.: Naming Infinity: A True Story of Religious Mysticism and Mathematical Creativity. The Belknap Press of Harvard Univ. Press (2009)

Graves, L.M., Hille, E., Smith, P.A., Zariski, O. (eds.): Proceedings of the International Congress of Mathematicians, Cambridge, Massachusetts, U.S.A., August 30-September 6, 1950. Amer. Math. Soc. (1952)

Greenaway, F.: Science International: A History of the International Council of Scientific Unions. Cambridge Univ. Press (1996)

Greenberg, D.S.: Space cooperation: U.S., Soviets agree to do up there what they have not done down here. Science 135(3509), 30 Mar, 1115–1117 (1962)

Greenberg, D.S.: The Smale Case: NSF and Berkeley pass through a case of jitters. Science 154(3745), 7 Oct, 130–133 (1966)

Gruber, C.S.: Academic freedom under pressure. Book review: 'No Ivory Tower' by Ellen W. Schrecker. Science 235(4786), 16 Jan, 371–372 (1987)

Guthrie, J.D.: The Second International Forestry Congress at Budapest. Science 84(2190), 18 Dec, 554–556 (1936)

Hamblin, J.D.: Visions of international scientific cooperation: the case of oceanic science, 1920–1955. Minerva **38**, 393–423 (2000a)

Hamblin, J.D.: Science in isolation: American marine geophysics research, 1950–1968. Physics in Perspective **2**(3), 293–312 (2000b)

Hargittai, B., Hargittai, I.: Dmitri I. Mendeleev: a centennial. Structural Chem. 18, 253–255 (2007)

Harvey, D.L., Ciccoritti, L.C.: US-Soviet cooperation in space. Center for Advanced International Studies, Univ. Miami (1974)

Hastings, A.B., Shimkin, M.B.: Medical research mission to the Soviet Union. Science 103(2681), 17 May, 605–608 (1946); ibid. 103(2682), 24 May, 637–644 (1946)

H.E.S.: Editor's note. Amer. Rev. Soviet Med. 1(1), 94 (1943)

Heymann, M., Martin-Nielsen, J.: Introduction: perspectives on Cold War science in small European states. Centaurus **55**(3), 221–242 (2013)

Hilton, R.: Russian and Soviet studies in France: teaching, research, libraries, archives, and publications. Russian Rev. **38**(1), 52–79 (1979)

Hodgson, P.E.: Conference on the Effects of Nuclear Weapons. Nature 176(4476), 13 Aug, 289–290 (1955)

Hoffman, P.: The Man Who Loved Only Numbers: The Story of Paul Erdős and the Search for Mathematical Truth. Hyperion (1998)

Holliday, R.: Conferences in Russia? Nature 245(5424), 12 Oct, 343 (1973)

Hollings, C.: The case of Evgenii Sergeevich Lyapin. Math. Today **48**(4), 184–186 (2012)

Hollings, C.: The struggle against idealism: Soviet ideology and mathematics. Notices Amer. Math. Soc. **60**(11), 1448–1458 (2013)

Hollings, C.: Mathematics across the Iron Curtain: A History of the Algebraic Theory of Semigroups. Amer. Math. Soc., Providence, Rhode Island (2014)

Hollings, C.: The acceptance of abstract algebra in the USSR, as viewed through periodic surveys of the progress of Soviet mathematical science. Historia Math. **42**(2), 193–222 (2015)

Houssay, B.A.: Twenty-First Congress, Buenos Aires—1959. In Fenn et al. (1968), pp. 50–54 (1968)

Howie, J.M.: Gordon Bamford Preston. Semigroup Forum **51**, 269–271 (1995)

Huxley, J.: A Scientist among the Soviets. Chatto and Windus, London (1932)

Ihde, A.J.: The Karlsruhe congress: a centennial retrospective. J. Chem. Ed. 38(2), 83–86 (1961)

Ingram, M., Roberts, T.A. (eds.): Botulism 1966: Proceedings of the Fifth International Symposium on Food Microbiology, Moscow, July 1966. Chapman & Hall, London (1967)

Ipatieff, V.N.: Modern science in Russia. Russian Rev. **2**(2), 68–80 (1943)

Isaev, A.N.: Soviet-French Seminar on Safety of Atomic Power Plants with Water-Moderated-Water-Cooled Reactors. Atomnaya energiya 46(5), 366–367 (1979) (in Russian); English trans.: Soviet Atomic Energy **46**(5), 424–426 (1979)

Ivanovskaya, A.A.: The origins of Soviet-French scientific-technical ties. Voprosy istorii, no. 1, 196–203 (1976) (in Russian)

Ivy, A.C.: The Fifteenth International Physiological Congress, Leningrad and Moscow, August 8–18, 1935. Amer. J. Digestive Diseases **2**(11), 692–695 (1935)

Joffe, A.: The main problems of physics in the U.S.S.R. Sci. Monthly 61(2), 154–156 (1945)

Johnston, T.: Being Soviet: Identity, Rumour, and Everyday Life under Stalin 1939–1953. Oxford Univ. Press (2011)

Joravsky, D.: The Lysenko Affair. Univ. Chicago Press (1970)

Josephson, P.R.: Soviet scientists and the state: politics, ideology, and fundamental research from Stalin to Gorbachev. Social Research **59**(3), 589–614 (1992)

Josephson, P.: Red Atom: Russia's Nuclear Power Program from Stalin to Today. Univ. Pittsburg Press (2000)

Kameneva, O.D.: Cultural rapprochement: the U.S.S.R. Society for Cultural Relations with Foreign Countries. Pacific Affairs **1**(5), 6–8 (1928)

Kamminga, H.: The International Union of Crystallography: its formation and early development. Acta Crystallogr. A **45**, 581–601 (1989)

Kassof, A.H.: Scholarly exchange and the collapse of communism. Soviet Post-Soviet Rev. **22**(3), 263–274 (1995)

Kato, G.: Twenty-Third Congress, Tokyo-1965. In Fenn et al. (1968), pp. 59–68 (1968)

Katz, M., Mac Lane, S., Adams, R.McC., Wilson, R.R.: Scientific exchanges with the Soviet Union. Bull. Amer. Acad. Arts Sci. **34**(1), 6–19 (1980)

Kautzleben, H., Müller, A.: Vladimir Ivanovich Vernadsky (1863–1945) — from mineral to noosphere. J. Geochem. Exploration **147**, 4–10 (2014)

Keltie, J.S., Mill, H.R. (eds.): Report of the Sixth International Geographical Congress, held in London, 1895. John Murray, London (1896)

Kerber, R.E.: A USA-USSR experiment in medical journalism: *The American Review of Soviet Medicine*. Amer. Communist Hist. **11**(2), 229–235 (2012)

Kevles, D.J.: 'Into hostile political camps': the reorganization of international science in World War I. Isis **62**(1), 47–60 (1971)

King, B.: SCR The Early Years. Anglo-Soviet J., Oct, 23–27 (1967)

Kirchik, O., Gingras, Y., Larivière, V.: Changes in publication languages and citation practices and their effect on the scientific impact of Russian science (1993–2010). J. Amer. Soc. Inform. Sci. Tech. **63**(7), 1411–1419 (2012)

Kirillov, P.L.: Soviet-French Seminar on Fast Reactors. Atomnaya energiya **42**(1), 64–65 (1977) (in Russian); English trans.: Soviet Atomic Energy **42**(1), 73–75 (1977)

Kline, J.R.: Soviet Mathematics. In: Christman, R.C. (ed.) Soviet Science: A Symposium Presented on December 27, 1951, at the Philadelphia Meeting of the American Association for the Advancement of Science, pp. 80–84. Amer. Assoc. Adv. Sci. (1952)

Knudsen, H., Nielsen, H.: Pursuing common cultural ideals: Niels Bohr, neutrality, and international scientific collaboration during the interwar period. In: Lettevall et al. (2012), pp. 115–139 (2012)

Kojevnikov, A.B.: Rockefeller philanthropy and Soviet science. Voprosy istor. estest. tekhn., no. 3, 80–111 (1993) (in Russian)

Kojevnikov, A.B.: Stalin's Great Science: The Times and Adventures of Soviet Physicists. Imperial College Press (2004)

Kolchinsky, E.I.: Nikolai Vavilov in the years of Stalin's 'Revolution from Above' (1929–1932). Centaurus **56**, 330–358 (2014)

Korneyev, S.C. (ed.): USSR Academy of Sciences: Scientific Relations with Great Britain. Nauka, Moscow (1977)

Korneyev, S.G., Timofeyev, I.A.: U.S.S.R. Academy of Sciences: relations with research institutions, scientists and scholars of Britain (1917–1975). In: Korneyev (1977), pp. 8–69 (1977)

Krementsov, N.A.: "Second front" in Soviet genetics: the international dimension of the Lysenko controversy, 1944–1947. J. Hist. Biol. **29**(2), 229–250 (1996)

Krementsov, N.: Russian science in the twentieth century. In: Krige and Pestre (1997), pp. 777–794

Krementsov, N.: The Cure: A Story of Cancer and Politics from the Annals of the Cold War. Univ. Chicago Press (2002)

Krementsov, N.: International Science between the World Wars: The Case of Genetics. Routledge, New York and London (2005)

Krementsov, N.: Big revolution, little revolution: science and politics in Bolshevik Russia. Social Research **73**(4), 1173–1204 (2006)

Krementsov, N.: In the shadow of the bomb: U.S.-Soviet biomedical relations in the early Cold War, 1944–1948. J. Cold War Stud. **9**(4), 41–67 (2007)

Krige, J.: NATO and the strengthening of Western science in the post-Sputnik era. Minerva **38**, 81–108 (2000)

Krige, J.: Atoms for peace, scientific internationalism, and scientific intelligence. Osiris **21**(1), 161–181 (2006)

Krige, J., Pestre, D. (eds.): Science in the Twentieth Century. Harwood Academic, Amsterdam (1997)

Kryuchkova, T.: English as a language of science in Russia. In: Ammon, U. (ed.) The Dominance of English as a Language of Science: Effects on Other Languages and Language Communities, pp. 405–423. Mouton de Gruyter, Berlin (2001)

Kurosh, A.G., Bityutskov, V.I., Boltyanskii, V.G., Dynkin, E.B., Shilov, G.E., Yushkevich, A.P. (eds.): Mathematics in the USSR after Forty Years, 1917–1957. 2 vols., Gos. Izdat. Fiz.-Mat. Lit., Moscow (1959) (in Russian)

Kutateladze, S.S.: Roots of Luzin's case. J. Appl. Ind. Math. **1**(3), 261–267 (2007)

Kutateladze, S.S.: The tragedy of mathematics in Russia. Siberian Electronic Math. Rep. **9**, A85–A100 (2012)

Kutateladze, S.S.: An epilog to the Luzin case. Siberian Electronic Math. Rep. **10**, A1–A6 (2013)

Kuznick, P.J.: Beyond the Laboratory: Scientists as Political Activists in 1930s America. Univ. Chicago Press (1987)

Langer, E.: Soviet genetics: first Russian visit since 1930's offers a glimpse. Science 157(3793), 8 Sept, 1153 (1967)

Lear, W.J.: Hot war creation, Cold War casualty: The American-Soviet Medical Society, 1943–1948. In: Fee, E., Brown, T.M. (eds.) Making Medical History: The Life and Times of Henry E. Sigerist, pp. 259–287. Johns Hopkins Univ. Press (1997)

Lefschetz, S.: The Second All-Soviet Mathematical Congress. Science 80(2082), 23 Nov, 479–480 (1934)

Lehto, O.: Mathematics without Borders: A History of the International Mathematical Union. Springer (1998)

Lettevall, R., Somsen, G., Widmalm, S. (eds.): Neutrality in Twentieth-Century Europe: Intersections of Science, Culture, and Politics after the First World War. Routledge, London and New York (2012)

Levering, R.B. (ed.): Debating the Origins of the Cold War: American and Russian Perspectives. Rowman & Littlefield (2002)

Levich, Y.: Trying to keep in touch. Nature 263(5576), 30 Sept, 366 (1976)

Levin, A.E.: Anatomy of a public campaign: "Academician Luzin's Case" in Soviet political history. Slavic Rev. **49**(1), 90–108 (1990)

Lorentz, G.G.: Who discovered analytic sets? Math. Intelligencer **23**(4), 28–32 (2001)

Lorentz, G.G.: Mathematics and politics in the Soviet Union from 1928 to 1953. J. Approx. Theory **116**, 169–223 (2002)

Lubrano, L.L.: National and international politics in US-USSR scientific cooperation. Social Stud. Sci. **11**(4), 451–480 (1981)

Lygo, E.: Promoting Soviet culture in Britain: the history of the Society for Cultural Relations between the Peoples of the British Commonwealth and the USSR, 1924–1945. Modern Lang. Rev. 108(2), 571–596 (2013)

Malkov, M.P. (ed.): Proceedings of the 10th International Conference on Low Temperature Physics: Moscow, USSR, 31 August – 6 September 1966. VINITI, Moscow (1967) (in Russian)

Manzione, J.: 'Amusing and amazing and practical and military': the legacy of scientific internationalism in American foreign policy, 1945–1963. Diplomatic Hist. **24**(1), 21–55 (2000)

Marton, E.: A portrait of Flóris Rómer in the frame of Budapest-Lisbon CIAAPS 1876–1880 congresses. In: Babes and Kaeser (2009), pp. 11–16 (2009)

Mazon, A.: Slavonic studies in France. Slavon. East Europ. Rev. **25**(64), 206–213 (1946)

Medvedev, Zh.A.: The Rise and Fall of T. D. Lysenko. Columbia Univ. Press (1969)

Medvedev, Zh.A.: The closed circuit - a record of Soviet scientific life. Nature 227(5264), 19 Sept, 1197–1202 (1970)

Medvedev, Zh.A.: The Medvedev Papers: The Plight of Soviet Science Today. Macmillan, London (1971)

Medvedev, Zh.A.: Soviet Science. Oxford Univ. Press (1979)

Medvedev, Zh.A.: East-West relations. Nature 310(5980), 30 Aug, 722 (1984)

Milanovsky, E.E.: Three sessions of the International Geological Congress held in Russia and the USSR (1897, 1937, 1984). Episodes: J. Internat. Geoscience 27(2), 101–106 (2004)

Miller, C.A.: 'An effective instrument of peace': scientific cooperation as an instrument of U.S. foreign policy, 1938–1950. Osiris 21(2), 133–160 (2006)

Nassau, J.J.: International relations in science and problems of visas. Science 124(3212), 20 Jul, 127 (1956)

Nathanson, M.B.: Math flows poorly from East to West. NY Times, 20 Sept (1986)

Needham, J., Davies, J.S. (eds.): Science in Soviet Russia. Watts and Co., London (1942)

Nemzer, L.: The Soviet Friendship Societies. Public Opinion Quarterly 13(2), 265–284 (1949)

Neswald, E.: Francis Gano Benedict's reports of visits to foreign laboratories and the Carnegie Nutrition Laboratory. Actes hist. cièn. tècn. 4, 11–32 (2011)

Neswald, E.: Strategies of international community-building in early twentieth century metabolism research: the foreign laboratory visits of Francis Gano Benedict. Hist. Stud. Nat. Sci. 43(1), 1–40 (2013)

Nielsen, H., Knudsen, H.: Too hot to handle: the controversial hunt for uranium in Greenland in the early Cold War. Centaurus 55(3), 319–343 (2013)

Nonweiler, T.: Seventh International Astronautical Congress. Nature 178(4538), 20 Oct, 832–833 (1956)

Novikov, Yu.M.: Soviet-French scientific-technical cooperation in the field of metrology. Izmeritel-naya tekhnika, no. 3, 80–81 (1973) (in Russian); English trans.: Measurement Techniques 16(3), 443–444 (1973)

Nygren, B.: The development of cooperation between the Soviet Union and three Western Great Powers, 1950–75. Cooperation and Conict 15(3), 117–140 (1980)

Oakes, M.E.: Science education and international understanding. Sci. Ed. 30(3), 136–148 (1946)

Palache, C.: The Geological Congress in Russia. Amer. Naturalist 31(371), 951–960 (1897)

Pechatnov, V.O.: The rise and fall of *Britansky Soyuznik*: a case study in Soviet response to British propaganda of the mid-1940s. Hist. J. 41(1), 293–301 (1998)

Pickavance, T.G., Skyrme, T.H.R.: High energy nuclear physics: Moscow conference. Nature 178(4525), 21 Jul, 115–116 (1956)

Plaud, R.: From the history of Franco-Russian and Franco-Soviet scientific ties. Voprosy istor. estest. tekhn., nos. 67–68, 91–98 (1980) (in Russian)

Raymond, E.A.: US-USSR cooperation in medicine and health. Russian Rev. 32(3), 229–240 (1973)

Redman, R.O.: International Astronomical Union: General Assembly in Dublin. Nature 176(4483), 1 Oct, 626–627 (1955)

Reid, M.: Keeping in touch with Soviet colleagues. Nature 265(5594), 10 Feb, 484–485 (1977)

Rich, V.: Russia's curtained window on the West. Nature 249(5457), 7 Jun, 502–504 (1974)

Rich, V.: Tbilisi: a conference with problems. Nature 257(5521), 4 Sept, 5–6 (1975a)

Rich, V.: Healthy exchanges. Nature 257(5526), 9 Oct, 441 (1975b)

Rich, V.: He who would dissident be. Nature 263(5576), 30 Sept, 361 (1976)

Rich, V.: US at sixes and sevens over Soviet exchanges. Nature 277(5697), 8 Feb, 424 (1979)

Richmond, Y.: Cultural Exchange and the Cold War: Raising the Iron Curtain. Pennsylvia State Univ. Press (2003)

Roberts, P.: Intelligence and internationalism: the Cold War career of Anton Bruun. Centaurus 55(3), 243–263 (2013)

Romanovsky, S.K.: Cultural and scientific ties between the Soviet Union and Great Britain. Anglo-Soviet J., Oct, 11–13 (1967)

Rosenzweig, M.R.: History of the International Union of Psychological Science (IUPsyS). Psychology Press, Hove (2000)

Rostov, S.: The professor who transplants hearts. Soviet War News, no. 1159, 18 May, 3 (1945)

Rudio, F. (ed.): Verhandlungen des ersten Internationalen Mathematiker-Kongresses in Zurich vom 9. bis 11. August 1897. Druck und Verlag von B. G. Teubner, Leipzig (1897)

Sagdeev, R.Z. (Eisenhower, S., ed.): The Making of a Soviet Scientist: My Adventures in Nuclear Fusion and Space from Stalin to Star Wars. Wiley (1994)

Sánchez-Ron, J.M.: International relations in Spanish physics from 1900 to the Cold War. Hist. Stud. Phys. Bio. Sci. **33**(1), 3–31 (2002)

Sapsai, A.: East-West relations. Nature 310(5977), 9 Aug, 446 (1984)

Schechter, B.: 'My Brain is Open': The Mathematical Journeys of Paul. Oxford Univ, Press. Oxford Univ. Press (1998)

Schrecker, E.W.: No Ivory Tower: McCarthyism and the Universities. Oxford Univ. Press. (1986)

Schrecker, E.: Many are the Crimes: McCarthyism in America. Princeton Univ. Press (1998)

Schroeder-Gudehus, B.: Challenge to transnational loyalties: international scientific organizations after the First World War. Sci. Stud. **3**(2), 93–118 (1973)

Schroeder-Gudehus, B.: Probing the master narrative of scientific internationalism: nationals and neutrals in the 1920s. In: Lettevall et al. (2012), pp. 19–42 (2012)

Schweitzer, G.E.: Techno-diplomacy: US–Soviet Confrontations in Science and Technology. Plenum Press, New York and London (1989)

Schweitzer, G.E.: Experiments in Cooperation: Assessing U.S.-Russian Programs in Science and Technology. The Twentieth Century Fund Press, New York. (1997)

Schweitzer, G.E.: Scientists, Engineers, and Track-Two Diplomacy: A Half-Century of U.S.-Russian Interacademy Cooperation. Nat. Acad. Press, Washington DC (2004)

Seligman, H., Russell, R.S.: Radioisotopes in scientific research: UNESCO conference. Nature 180(4594), 16 Nov, 1029–1031 (1957)

Semenov, B.A.: Soviet-French collaboration in the field of the peaceful utilisation of atomic energy. Atomnaya energiya **46**(3), 201–203 (1979) (in Russian); English trans.: Soviet Atomic Energy **46**(3), 236–238 (1979)

Shapley, D.: Détente: travel curbs hinder U.S.-U.S.S.R. exchanges. Science **186**(4165), 22 Nov, 712–715 (1974)

Shishkin, B.: The work of Soviet botanists. Science 97(2520), 16 Apr, 354–355 (1943)

Siegmund-Schultze, R.: Rockefeller and the Internationalization of Mathematics between the Two World Wars: Documents and Studies for the Social History of Mathematics in the 20th Century. Birkhäuser Verlag, Basel, Boston (2001)

Siegmund-Schultze, R.: One hundred years after the Great War (1914–2014): a century of breakdowns, resumptions and fundamental changes in international mathematical communication. In: Jang, S.Y., Kim, Y.R., Lee, D.-W., Yie, I. (eds.) Proceedings of the International Congress of Mathematicians Seoul 2014, vol. IV, pp. 1231–1253. Gyeong Munsa (2014)

Sigerist, H.E.: Editorial: on American-Soviet medical relations. Amer. Rev. Soviet Med. **5**(1), 4–8 (1948a)

Sigerist, H.E.: Editorial. Amer. Rev. Soviet Med. **5**(4), 162 (1948b)

Smale, S.: On the steps of Moscow University. Math. Intelligencer **6**(2), 21–27 (1984)

Smith, M.B.: Peaceful coexistence at all costs: Cold War exchanges between Britain and the Soviet Union in 1956. Cold War Hist. **12**(3), 537–558 (2012)

Smith, R.A.: Physics of semiconductors. Nature 188(4751), 19 Nov, 632–633 (1960)

Solomon, S.G., Krementsov, N.: Giving and taking across borders: the Rockefeller Foundation and Russia, 1919–1928. Minerva **39**, 265–298 (2001)

Sommer, U.: The International Congress of Prehistoric Anthropology and Archaeology and German Archaeology. In: Babes and Kaeser (2009), pp. 17–31 (2009)

Sorokina, M.: Partners of choice/*Faute de mieux*? Russians and Germans at the 200th anniversary of the Academy of Sciences, 1925. In: Solomon, S.G. (ed.) Doing Medicine Together: Germany and Russia Between the Wars, pp. 61–102. Univ. Toronto Press (2006)

Stern, L.: The background history of creation of the French Rapprochement Society: The New Russia (based on the unpublished VOKS documents). Austral. Slavon. East Europ. Stud. 11(1–2), 143–160 (1997) (in Russian)

Stern, L.: The All-Union Society for Cultural Relations with Foreign Countries and French Intellectuals, 1925–29. Austral. J. Politics Hist. 45(1), 99–109 (1999)

Strakhovsky, L.I.: On understanding Russia: a review article. Amer. Slavic East Europ. Rev. **6**(1/2), 181–199 (1947)

Strekopytov, S.P.: International scientifico-technical connections in the first years of Soviet power. Voprosy istor. estest. tekhn., no. 2(59), 74–78 (1977) (in Russian)

Struve, O.: The International Astronomical Union. Science 117(3039), 27 Mar, 315–318 (1953)

Sweeney, C.: Soviet scientist is made an exile. Times (London), 9 Aug, 5 (1973)

Todd, J.: The SCR 1937–1952. Anglo-Soviet J., Oct, 28–32 (1967)

Todes, D., Krementsov, N.: Dialectical materialism and Soviet science in the 1920s and 1930s. In: Leatherbarrow, W.J., Offord, D. (eds.) A History of Russian Thought, pp. 340–367. Cambridge Univ. Press (2010)

Tokareva, T.A.: The first congresses of domestic mathematicians: the prehistory and formation of the Soviet mathematical school. Istor.-mat. issled., no. 6(41), 213–231 (2001) (in Russian)

Topchiev, A.V.: The USSR Academy of Sciences and its scientific links with Britain. Anglo-Soviet J., Mar, 26–29 (1956)

Trostnikov, V.N.: Worldwide Congress of Mathematicians in Moscow. Izdat. Znanie, Moscow (1967) (in Russian)

Tucker, A.W.: The Topological Congress in Moscow. Bull. Amer. Math. Soc. **4**(11), 764 (1935)

Villat, H. (ed.): Comptes rendus du Congrès international des mathématiciens (Strasbourg, 22–30 Septembre 1920). Imprimerie et Librairie Édouard Privat, Toulouse (1921)

Vinogradov, I.: Mathematics in the U.S.S.R. Nature 150(3815), 12 Dec, 677–678 (1942)

Vucinich, A.: Science in Russian Culture 1861–1917. Stanford Univ. Press (1970)

Vucinich, A.: Mathematics and dialectics in the Soviet Union: the pre-Stalin period. Historia Math. 26, 107–124 (1999)

Vucinich, A.: Soviet mathematics and dialectics in the Stalin era. Historia Math. **27**, 54–76 (2000)

Vucinich, A.: Soviet mathematics and dialectics in the post-Stalin era: new horizons. Historia Math. **29**, 13–39 (2002)

Waddington, C.H.: International Genetics Symposia in Japan. Nature 178(4546),15 Dec, 1329–1330 (1956)

Wang, Z.: U.S.-China scientific exchange: a case study of state-sponsored scientific international-ism during the Cold War and beyond. Hist. Studies Phys. Biol. Sci. 30(1), 30, 249–277 (1999)

Watson-Jones, R.: Russian surgeons and Russian surgery. Bull. War Med. **4**(3), 121–123 (1943a)

Watson-Jones, R.: Surgical mission to Moscow. Brit. Med. J. **2**(4312), 28th August, 276–277 (1943b)

Webb-Johnson, A.E.: Surgical instruments for Russia. Brit. Med. J. **2**(4222), 6 Dec, 830 (1941); ibid. **1**(4235), 7 Mar, 335 (1942); Lancet **239**(6183), 28 Feb, 275 (1942)

Whitney, H.: Moscow 1935: topology moving toward America. In: Duren, P.L. (ed.) A Century of Mathematics in America, vol. 1, pp. 97–117. Amer. Math. Soc. (1989)

Wiley, H.W.: Second International Congress of Applied Chemistry. J. Amer. Chem. Soc. **18**, 923–940 (1896)

Wolfe, A.J.: Competing with the Soviets: Science, Technology, and the State in Cold War America. Johns Hopkins Univ. Press (2013)

Wooster, W.A.: Order-disorder structures. Nature 202(4393), 27 Jun, 1273–1274 (1964)

Wurtz, C.-A.: Account of the sessions of the International Congress of Chemists in Karlsruhe, on 3, 4, and 5 September 1860. In Kekulé, A. (ed.) Richard Anschütz, vol. 1, pp. 671–688. Verlag Chemie, Berlin (1929); English trans. in Nye, M.J.: The Question of the Atom. Tomash, Los Angeles (1984)

Yushkevich, A.P.: The case of Academician N. N. Luzin. Vremya. Idei. Sudby, 12 Apr (1989) (in Russian)

Zavyalskii, L.P.: Soviet-French Symposium on Fuel Elements of Fast Reactors. Atomnaya energiya 35(1), 73–75 (1973) (in Russian); English trans.: Soviet Atomic Energy **35**(1), 695–696 (1973)

Zhebrak, A.R.: Soviet biology. Science 102(2649), 5 Oct, 357–358 (1945)

Zhukovsky [sic]: The cosmic rays mystery. Soviet War News, no. 1313, 29 Nov, 3 (1945)

Ziman, J.: Letter to an imaginary Soviet scientist. Nature 217(5124), 13 Jan, 123–124 (1968)

Chapter 3
Physical Access to Publications

Abstract In this chapter, we address issues surrounding physical access to publications. We consider the matter of censorship, and the tendency of Soviet scientists to publish much of their work in 'local' journals, which often did not find their way into Western libraries. We deal with efforts to gain broad impressions of the work of 'the other side': for instance, abstracting services, and the many published Western surveys of Soviet scientific advances.

Keywords Library access to publications · Censorship · 'Local publication' in the USSR · Abstracts · Western surveys of Soviet science

3.1 General Comments

So far, we have confined our attention to *personal* contacts between scientists in East and West, so we turn now to an aspect of international scientific communication that might be characterised as being rather more *impersonal*: the exchange of publications. It was of course through published matter that those scientists who were not able to communicate directly with their opposite numbers were able, at least in principle, to keep abreast of international developments in their field. As noted in Chap. 1, there are two broad dimensions to this issue: the question of *physical* access to materials, and that of *linguistic* access. The latter is addressed in the next chapter, but it is the former that concerns us here.

Alongside the comments in Sect. 2.1 concerning the visibility of Russian researchers on the international scientific scene in the decades prior to the First World War, we might also mention that Russian publications appear to have enjoyed a certain circulation outside Russia at this time; indeed, the world-renowned Russian scientists who were mentioned in Sect. 2.1 could not have built international reputations on conference papers alone. As we will discuss in Chap. 4, many Russian publications were in foreign languages, and so were evidently written with an international audience in mind, but it seems that even some periodicals that appeared entirely in Russian nevertheless enjoyed a foreign readership. We might take here the oft-cited example of the journal *Matematicheskii sbornik*. As we will see in

C.D. Hollings, *Scientific Communication Across the Iron Curtain*,
SpringerBriefs in History of Science and Technology,
DOI 10.1007/978-3-319-25346-6_3

Sect. 4.2, this is a journal that was published entirely in Russian from its foundation in 1866 until a change of language policy in 1922. However, if we look to
the early volumes of the 20th century, we find lists of subscribers to the journal,
including many, both individual and institutional, in foreign countries. Volume 21
(1900–1901), for instance, was shipped to various cities across France, Germany,
Austria-Hungary, Sweden, Spain, the USA, Great Britain, the Netherlands, Italy,
Norway, and Portugal (Anon 1901b, p. 785). Similar lists appear in other volumes
of the journal from the late-19th and early-20th centuries.[1] Together with these, we
often find further lists of foreign materials that had been received by the Moscow
Mathematical Society that year. To take the example of volume 21 again, we note
that the Society had, in the relevant period (late 1900/early 1901), received periodicals from Sweden, Finland, France, Argentina, the USA, Spain, Austria-Hungary,
Germany, Portugal, the Netherlands, and Italy (Anon 1901a, pp. 780–781).

Moving beyond the October Revolution, we note that, in contrast to what seems
to be the popular view on this issue, the physical access enjoyed by scientists on one
side of what became the Iron Curtain to the publications of those on the other was,
generally speaking, quite good throughout the 20th century. Indeed, if we look carefully at the general Western scientific literature, we do find the occasional acknowledgement of this.[2] Naturally, however, these statements require a certain amount
of qualification. There were, for example, particular periods when this access was
disrupted: most notably, the two world wars. During wartime, the interruption of
the flow of information was one of great concern, and so, in 1916, discussions were
held concerning the possible formation of an international scientific organisation
that would facilitate the exchange of scientific materials between Russia, France
and Great Britain (Menschutkin 1917), although these talks do not appear to have
come to anything—probably because the October Revolution intervened. Nevertheless, during the world wars, and also the period of confusion in Russia that followed
the revolution and civil war, we find frequent appeals from one side for the literature of the other.[3] Thus, like the scope for personal communications, access to published materials varied over the decades. Moreover, an individual scientist's access
to foreign publications depended very much upon where he or she was based: major
academic centres such as Oxford, Paris, Moscow, etc. typically had better provision
than less prominent establishments. The point of origin of publications is also a factor to consider: the journals produced by major, centrally-based, publishers, such as
the Soviet Academy of Sciences were more likely to be found in libraries on the
opposite side of the Iron Curtain.

Various national and local organisations had a large role to play in the exchange
of publications. To take an early example, the first years of the 1920s saw the
formation of the British Committee for Aiding Men of Letters and Science in

---

[1] See also the comments of Demidov (1996, p. 136).

[2] Kline (1952, p. 83) remarked, for example, that "[i]t is possible to secure Russian mathematical
journals with comparative ease"; similar remarks concerning the Soviet archaeology literature may
also be found in Chard (1969, p. 774).

[3] See, for example, Kellogg (1922), Razran (1942), and Anon (1943).

Russia[4]; simply put, the goal of this body was the organised mailing of British literary and scientific works to Russian academics who might not otherwise have been able to obtain them under the then-uncertain political conditions. Similar bodies were also set up to supply the scholars of the fledgling Soviet Union with US, German, and French materials.[5] Although efforts by Lenin in the early 1920s to establish exchanges between the Petrograd public library and US, British and German scientific institutions had proved unsuccessful (Gak 1963, pp. 196–197), VOKS subsequently became very active in the purchase of foreign materials: in 1934 it imported 163,000 volumes from 84 countries, much of which was scientific literature; these materials were distributed to over 1,500 Soviet institutions (David-Fox 2012, p. 299). In addition, two-way exchanges were established in the 1920s, both by VOKS (Kameneva 1928, pp. 7–8) and by other organisations: for example, there was a book exchange programme between the Soviet Academy of Sciences and the Smithsonian Institution in the USA (Furaev 1974, English trans., p. 57). Many similar such two-way exchanges waxed and waned over the decades (Arutjunov 1979). It should be noted that although it became possible in later years for Soviet scientists to take out *personal* subscriptions to Western periodicals (subject, however, to some of the problems of censorship that are to be discussed in Sect. 3.2),[6] *institutional* subscriptions remained the norm, as is the case in the West.

During the later years of our period of interest (specifically, from 1952 onwards), the import into, and distribution of foreign scientific publications within, the USSR was handled centrally by a specially-established branch of the Academy of Sciences: the All-Union Institute for Scientific and Technical Information, or VINITI (ВИНИТИ = Всесоюзный институт научной и технической информации; now the All-*Russian* Institute for Scientific and Technical Information: see Panov 1956). VINITI was also responsible for the publication of a monthly calendar of upcoming international conferences in all areas of science (Medvedev 1971, p. 128), and of an abstracting journal, *Referativnyi zhurnal* (*Реферативный журнал*), about which I will say more in Sect. 3.4. The range of materials obtained by VINITI appears to have been quite broad, covering (in line with the comments made above) both national Western publications, such as the *Proceedings of the National Academy of Sciences of the USA* (Medvedev 1971, p. 45), and also periodicals of a more 'local' nature, such as the *Proceedings of the Cambridge Philosophical Society* (Hollings 2014, p. 30). There were, however, certain delays in the supply of Western sources to Soviet scientists, caused, in part, by censorship (see Sect. 3.2), and also simply by the way in which VINITI operated. For reasons of economy, only very few copies of an issue of a journal were purchased. These were then reproduced, and copies sent to libraries across the USSR.[7]

---

[4] See Montagu et al. (1921), Schuster (1921), Anon (1921), and Gregory and Wright (1922).

[5] See Solomon and Krementsov (2001, pp. 271, 277, 287). On efforts to supply Soviet scientists with German publications, see also Forman (1973, p. 167).

[6] Zhores Medvedev, for example, subscribed to *Science*: see Medvedev (1971, p. 343).

[7] See Medvedev (1971, pp. 124–125, 361–362) or Rich (1974, p. 504).

It should be noted in this connection that the USSR did not join the Universal Copyright Convention until 1973 (Bloom 1973; Garfield 1973).

In the West, the procurement of Soviet scientific literature was much less centralised. In the United States, for instance, the Library of Congress and the Foreign Technical Information Center of the Department of Commerce both sought out Soviet (scientific) literature (Sherrod 1958; Anon 1958). As a guide to materials obtained, the Library of Congress published a *Monthly Index of Russian Accessions*; it was estimated that in 1958 the library was acquiring 60 % of Soviet publications in the natural sciences (Sherrod 1958, p. 958). The US National Science Foundation provided a great deal of funding for the purchase (and, indeed, translation—see Sect. 4.5) of Soviet materials (Anon 1959a). In the UK, both the British Library and the Bodleian Library in Oxford, for example, maintain a wealth of material published in the Soviet Union, though the occasional gaps in their collections may indicate that the acquisition of these did not always run smoothly (see Sect. 3.3). Avenues for *individuals* to obtain Soviet scientific materials were also open (Friedman 1967).

As already noted in Chap. 1, lack of Western knowledge of Soviet scientific research, and (less frequently, as I will argue) vice versa, led to the duplication of some work, and to the occasional accusation of plagiarism, or to charges that Western scientists did not cite Soviet work as often as they perhaps ought to (Anon 1982). However, this was not, I contend, because of a lack of Soviet literature in the West: the materials were available, but, as I will discuss in Chap. 4, they were in a language that a large number of Western readers could not decipher. Connected with this, there also seems to have been a certain lack of knowledge on the part of Western readers about the Soviet resources that were available to them. There thus appeared several guides to the Russian holdings of many Western libraries[8]—we will see in Sect. 4.5 that similar guides were produced in response to much the same problem in connection with the provision of translations. We even find guides (published in the West) to the foreign holdings of Soviet libraries,[9] though one wonders whether such guides would have had a wide circulation within the USSR—as we might expect, foreign materials were handled very cautiously by Soviet librarians, and were often subject to varying degrees of censorship.

3.2 Censorship

As I have already indicated in Sect. 2.7, censorship was an issue faced by Soviet scientists. Alongside the postal censorship suspected by Zhores Medvedev, there also occurred an organised programme of expurgation of imported materials. Western publications entering the USSR were stripped of any material deemed to be politically sensitive, including, for example, any references to the work of 'polit-

---

[8]In the case of the US Library of Congress, we have, for example, Horecky (1964) and Kraus (1976, 1979).

[9]See, for example, Kuhterina (1980).

ically undesirable' Soviet figures, such as refuseniks (Rich 1975). *The Medvedev papers* has a great deal to say, for example, on the "complicated surgery" undergone by issues of the journal *Science* (Medvedev 1971, p. 356), and also the efforts that Soviet censors made in order to disguise their handiwork: by, for example, replacing deleted articles by advertisements from earlier issues of a journal, in order to preserve the pagination.[10] This treatment, moreover did not escape the notice of the journal's editors (Carey 1983). Articles removed by Soviet censors included a piece concerning the preparation of journal offprints—something with which Soviet authors were rarely supplied.[11] In those instances when the exchange of offprints between Western and Soviet scientists was possible, it was handled centrally by VOKS (Krementsov 2005, pp. 42–43).

Occasional concerns about postal censorship were also raised in the West. For example, in January 1962, the US House of Representatives passed the so-called Cunningham Amendment, a measure designed to prevent the circulation of communist propaganda within the US postal system (Anon 1962a). US academics who took a scholarly interest in the principles and practice of communism began to worry that materials necessary for their work would be denied to them. Indeed, the problem had the potential to be wider still, given the Soviet habit of including articles of an ideological nature in journals (scientific journals in particular) where, it might reasonably be argued, such material has no place.[12] However, Congressman Cunningham's reassurances that scholarly work would not be affected do in fact appear to have been genuine (Anon 1962a, p. 16)—I have found no particular evidence of interference by the US postal service in the distribution of scientific materials from the USSR.

Even if fears of censorship of materials *entering* the USA proved to be unfounded, there were nevertheless restrictions on the material that might be exported: depending on the subject-area, US scientists were sometimes required to seek permission before submitting papers to international journals (Hamblin 2000, p. 296). This appears to have been a particular cause of concern in the early 1980s (see, for example, Panel 1982): ideals of scientific freedom clashed with the worries of the US Department of Defence (DoD) that certain papers delivered by US delegates at international scientific conferences were communicating sensitive information directly to any Soviets who happened to be in attendance. In a much-publicised incident, around 100 papers by US speakers were withdrawn from the Twenty-Sixth Annual International Technical Symposium of the Society of Photo-Optical Instrumentation Engineers, held in San Diego, California, in August 1982, after US officials warned that some talks were too sensitive to be delivered, in light of the fact that 28 Soviet delegates were expected at the conference.[13] In the aftermath, a great deal of ink was

---

[10]See Medvedev (1971, p. 360) or Hollings (2014, pp. 30–31).

[11]Medvedev (1971, p. 124). Even when they had offprints to send, some Soviet scientists would probably have thought twice about doing so, for fear of being accused of 'collaborating with the enemy': see Josephson (1992, p. 598).

[12]See, for example, various pieces in vol. 70, no. 2 (1970) of *Uspekhi matematicheskii nauk*: Anon (1970), Gnedenko (1970), and Lapko and Lyusternik (1970).

[13]See Greenberg (1982), Boffey (1982), Shapley (1982), and Kolata (1982).

expended on the needs of national security versus the right to unhindered scientific communication.[14] However, the curtailment of scientific contacts by the DoD was not as comprehensive as that practiced in the USSR (which occasionally banned the export of whole journals: see, for example, Schwartz 1951), and, indeed, negotiations were possible between the DoD and US scientific organisations (Peterson 1982b).

3.3 'Local Publication' in the USSR

As noted in Sect. 3.1, the provision of Soviet publications in the West depended very much, as one might expect, on the point of origin of the particular book or journal. Those journals produced by the Academy of Sciences, for example, had a wide circulation outside the USSR. Moreover, those materials that originated in the Soviet Union's 'academic core' of Moscow/Leningrad were, generally speaking, more likely to find their way into Western libraries than those from elsewhere in the Soviet Union. We have already encountered, for example, the journal *Matematicheskii sbornik* (in Sect. 2.3), published in Moscow. As a brief scan of the electronic catalogues of major Western (academic) libraries reveals, this journal has traditionally been quite widely available in the West.[15] There are, however, exceptions to this loosely-formulated 'Moscow/Leningrad rule': to cite a personal example, I have experienced some difficulty in accessing (in British libraries) the publications of the Leningrad State Pedagogical Institute (now the Herzen Russian State Pedagogical University)—difficulties that I have not usually encountered with those of Leningrad State University, for instance.

The provision in Western university libraries of Soviet materials published outside Moscow and Leningrad is much more patchy. Again, a perusal of electronic catalogues will bear this out. To take one example, the Radcliffe Science Library in Oxford holds only the first seven volumes (1935–1939) of the Ukrainian journal *Учені записки Харківського державного університету* (*Scientific Notes of Kharkov State University*),[16] which ran under this name until its 145th volume in 1964. Whether the Oxford library's failure to maintain its subscription to this journal was a result of difficulties in obtaining it, or simply the cancellation of a little-used resource (volumes 1–23 of the journal were published in Ukrainian—perhaps not the most accessible language for the typical British reader), must remain a matter for speculation. Nevertheless, this example serves as a single illustration of the incomplete nature of Western library holdings of Soviet materials—there are many others.

---

[14]See, for example, Peterson (1982a, c) and Park (1986).

[15]Indeed, the journal is now more widely available in the West than ever, with all back issues freely accessible (along with those of many other former Soviet mathematical journals) on the site 'Math-Net.ru', on which, see Chebukov et al. (2013).

[16]In Russian (under which title this journal is often cited): *Ученые записки Харьковского государственного университета.*

The hit-and-miss character of Western library provision of Soviet resources from outside Moscow and Leningrad is particularly unfortunate when we note that Soviet scientists appear, in many instances, to have been encouraged to publish their work in 'local journals' from around the mid-1930s. By 'local journals', I mean those published by a particular scientist's own institution, rather than 'national' publications, like those produced by the Academy of Sciences. Though Soviet scientists did not cease to publish in the latter, the bulk of their published work appears, in many cases, to have been confined to their universities' periodicals. My claim concerning the drive towards 'local publication' is, unfortunately, an impressionistic one: I do not, at present, have any evidence to offer other than my own anecdotal estimations, formed after perusing the lists of publications of many Soviet scientists. In particular, I have yet to find any explicit acknowledgement or explanation of this trend in the Soviet scientific literature. I offer some limited speculation on the motivation for the shift: it may simply have been easier, bureaucratically speaking, for scientists to publish their work locally, or it may perhaps have been an overreaction to the injunction against foreign publication (as discussed in Sect. 2.3).

Whatever its causes, the movement towards 'local publication' was observed, with perhaps a hint of frustration, by Western commentators. Take, for example, the following passage from a 1962 appraisal of Soviet mathematics:

> In the Soviet Union ... an important paper may turn up in the *Uchenye Zapiski* [*Scientific Notes*] of a small pedagogical institute in Ulan-Ude or Irkutsk, buried among less noteworthy writings in the broad scientific field, and it may never be available outside the USSR. (Anon 1962b, p. 13)

In fact, to pursue these references to Ulan-Ude and Irkutsk, both of which are in Siberia, we note that there is one further Siberian example, from which many scientific publications do appear to have reached the West, namely Novosibirsk. As an example from mathematics, we have the journal *Алгебра и логика* (*Algebra and Logic*), founded in Novosibirsk in 1962, and published by the Siberian branch of the Academy of Sciences. Taken collectively, the holdings for this journal in British university libraries are reasonably complete. Indeed, the journal's connection to the Academy of Sciences, together with the concomitant prestige, is probably the reason that it found its way into Western libraries. Other Novosibirsk-based journals are similarly well-represented: the various series of *Известия Сибирского отделения Академии наук СССР* (*Bulletin of the Siberian Branch of the Academy of Sciences of the USSR*) enjoy broad (though by no means complete) holdings in, for example, the Radcliffe Science Library in Oxford.

3.4 Abstracts

For a time, the comprehension of Russian papers by Western readers was aided by the fact that some Soviet journals carried abstracts in languages other than Russian. This is something that I will deal with in Sect. 4.2. For the time being, however, we note that Western readers were also helped out by a slightly different type of abstracts: those found in abstracting journals. Whether or not a given Soviet resource was widely available in the West, a general familiarity with the Soviet scientific literature was fostered through the coverage of Russian journals by the various regular Western abstracting publications,[17] such as (to cite some English-language ones) the *Chemical Abstracts Service* (see Baker et al. 1980), *Biological Abstracts* (Sinclair 1953), *Physics Abstracts* (Vlachý 1979), *Psychological Abstracts* (Fernberger 1938; Benjamin and VandenBos 2006), and *Mathematical Reviews* (Lehmer 1989; Pitcher 1988; Richert 2014). Indeed, to take the latter as a case in point, we note that when this abstracting journal was established in 1940, one of the express goals of the founders was to review papers published in as many languages as possible, thereby enabling researchers at least to become acquainted with the broad strokes of materials that they might not otherwise have been able to read. Three of *Mathematical Reviews*' sometime-editors, namely S.H. Gould (1956–1962), A.J. Lohwater (1961–1965) and J. Burlak (1971–1977), probably had a hand in the treatment of Soviet sources, since all three subsequently compiled Russian-English mathematical dictionaries and language guides.[18] To remain with mathematics for the moment, we note also that the German abstracting journal *Zentralblatt für Mathematik*, which became an East/West co-publication following the division of Germany, later proved to be a useful conduit for Western understanding of Soviet mathematical work since East German mathematicians who had learnt Russian at school were able to produce accessible German reviews of Russian papers (Siegmund-Schultze 2014, pp. 1245–1246).

Extensive abstracting activities also took place on the opposite side of the Iron Curtain (Panov 1955; Beyerly 1956; Gordin 2015, pp. 248–251). Prior to the Second World War, several separate abstracting journals operated. One of the first of these was Центральный реферативный медицинский журнал (*Central Medical Abstracting Journal*), founded in 1928; amongst the others was, for example, Физико-математический журнал (*Physico-Mathematical Journal*). However, the USSR's entry into the Second World War marked the end of publication for these periodicals—widespread abstraction of both domestic and foreign scientific publications did not recommence in the USSR until 1953, with the launch by VINITI of the major abstracting journal *Referativnyi zhurnal*. In its early years, *Referativnyi zhurnal* was published in eight sections (chemistry,[19] biology, physics, astronomy, geology and geography, mathematics, mechanics, and biochemistry) but several new sections were subsequently added. Like those of the Western abstracting services, the editors of *Referativnyi zhurnal* endeavoured to cover materials published in a range of languages—one count suggested that abstracts of papers

---

[17]There were also irregular and occasional abstracting services, such as those provided by *The American Review of Soviet Medicine* (see Sect. 2.4), those attempted by the SCR (Lygo 2013, pp. 589–590), and, somewhat earlier, the occasional abstracts of relevant Russian papers that were produced by the US Department of Agriculture (Benedict 1909).

[18]See note 40 on p. 87.

[19]On the chemical section of *Referativnyi zhurnal*, see Serpinsky (1956).

written in as many as 34 (mostly European) languages found their way into the pages of *Referativnyi zhurnal* (Beyerly 1956, p. 137). One point upon which this Soviet abstracting journal differed from many of its Western counterparts, however, was in its evaluation of the papers under review: whereas a neutral and objective point of view is typically (though not universally) adopted by Western abstractors, usually resulting in purely descriptive reviews,[20] Soviet abstracts often featured critical (even ideological) evaluations (Beyerly 1956, pp. 138–139).

Besides providing readers with a summary of papers that were of potential use to them, abstracting services sometimes offered a photocopying scheme, whereby readers could purchase copies of materials in which they were interested. This was the case with *Referatvnyi zhurnal*, for example.[21] Upon its foundation in 1940, *Mathematical Reviews* sold both photo- and microfilm copies of papers that it had reviewed (but not of books or copyrighted material), although this service was discontinued in 1947, owing to the practicalities of handling the enormous amount of material that was by then available (Pitcher 1988, pp. 72–73).

As can be seen from some of the sources already cited,[22] the extensive abstracting activities in the USSR did not escape Western notice. Indeed, the centralised and comprehensive nature of Soviet abstracting became a source of concern to some Western commentators. In an article in *The New York Times* in May 1958, one such author wrote that

> [a] Soviet biologist somewhere in Siberia could snuggle up in bed and read comfortably about the whistling habits of bobwhites in Iowa, the annual report of the Calcutta inland fisheries research station, a Japanese study of the effects of radioactivity in Bikini waters and the isolation of a rabies virus from native Ohio bats. (Frankel 1958)

The feeling appears to have emerged in the USA in particular that US (more generally, Western) scientific developments were more widely available to the scientists of the USSR than was the case in the opposite direction.[23] As I will argue more fully in Chap. 4, this greater availability stemmed from the language skills of Soviet scientists: as I indicated in Sect. 3.1, a very broad selection of the scientific works of the USSR appears to have been (physically) available in the West, but this went largely unread, thanks to the language barrier. Nevertheless, a fear of Soviet technological advances, coupled with simple curiosity, led to a desire in the West to go beyond mere abstracts of papers, and to learn a great deal more about Soviet scientific research.

---

[20]The current *Mathematical Reviews* 'Guide for Reviewers' states, for example, that "[a] review should primarily help the reader decide whether or not to read the original item", and that "critical remarks should be objective, precise, documented and expressed in good taste[: v]ague criticism offends authors and fails to enlighten the reader" (http://www.ams.org/mresubs/guide-reviewers. html — last accessed 26th May 2015).

[21]See Beyerly (1956, p. 139) or Tareev (1962, p. 341).

[22]For example, Beyerly (1956); see also DuS (1956).

[23]Similar fears had been expressed decades earlier concerning German science: that extensive German abstracting activities might lead to German scientific dominance, even in the wake of the defeat of the First World War (Siegmund-Schultze 1994, pp. 306–307, 311).

3.5 Western Surveys of Soviet Work

In the decades following the Second World War there emerged a quite natural desire amongst Western scientists not only to learn more about the specific details of Soviet science, but also to obtain a broader view of the general themes, and, indeed, to gain a more comprehensive impression of Soviet scientific culture. In many instances, this was motivated by curiosity (and perhaps a desire not to duplicate Soviet work), but in others, it may have owed more to political considerations, and a fear of 'the other side'. Indeed, this was a fear that was exacerbated by the surprise Soviet launch of Sputnik I in October 1957. As Warren B. Walsh commented in 1960:

> The pained and slightly incredulous astonishment with which we have reacted to repeated, spectacular demonstrations of Soviet scientific and technical prowess is partially a reflection of past inattention and ignorance. (Walsh 1960, p. 277)

The need to learn about Soviet advances thus became imperative, and was driven not only by the "almost irrational howl of horror" from the US press (Cadbury 2005, p. 168), but also by claims that the launch of Sputnik had been foreshadowed in Russian scientific literature (Beyer 1965, p. 46). Whether or not this last assertion was true, it led to the belief that Western scientists could have been prepared for Sputnik, had they been more familiar with Soviet scientific publications. I will say a little more about Sputnik's impact in Sects. 4.4 and 4.5.

The general Western scientific (indeed, academic) literature contains many accounts by researchers of visits made to the USSR,[24] featuring detailed observations on the status of their particular field, and also more general impressions of life in the Soviet Union. I have already cited some reports of this type, in connection with wartime medical exchanges (Sect. 2.4), and, indeed, international conferences that were held in the USSR both before and after the Second World War: take, for example, the account by W. Bateson of his visit to the USSR in 1925 (Bateson 1925). Moreover, I have mentioned, for instance, the book written by Julian Huxley following his visit to the USSR in the Summer of 1931 (Huxley 1932). Such accounts are merely the scientific manifestation of the more general efforts, which had been underway since the early 1920s, and often had their origins in visits to the USSR, to engender greater understanding of the Soviet Union in the West.

Reports of foreign trips were also produced by Soviet visitors to the West; these were usually for the benefit of the relevant Soviet government agency,[25] or for the Overseas Section of the Academy of Sciences (Levich 1976, p. 366), but they were occasionally published for general consumption (Medvedev 1971, pp. 118–119): see, for example, the Soviet accounts of the Edinburgh ICM of 1958.[26] Another published Soviet report, which nowadays makes rather entertaining reading, is that

---

[24]Such as Ashby (1947), Gerard (1950), Anon (1954), Penfield (1955), Piaget (1956), Lohwater (1957), Bockris (1958), Anon (1959b), Armstrong (1961), Anon (1961), Charlier and Dietz (1966), Thwaites (1968), Abelson (1966), and Danckwerts (1983).

[25]See Richmond (2003, p. 73) or Gerovitch (2002, p. 156).

[26]See vol. 14, no. 2 of *Uspekhi matematicheskikh nauk* for 1959 for reports both of a general character, and also on the treatment of specific branches of mathematics at the congress.

on the Eighteenth International Physiological Congress in Copenhagen in 1950 (see Sect. 2.7). The report was published first in *Izvestiya*, and then reproduced (in English translation) in *Science* (Bykov 1950). Its author, Academician C. Bykov, made it quite clear to his Soviet readers that this conference "was not arranged as well as the 15th Congress which took place here in Moscow and Leningrad in 1935" (see Sect. 2.3), and noted that the arrival of the Soviet delegation was accompanied by "cheap and noisy sensationalism of the American style" (Bykov 1950, p. 768). The "aggressive tendencies of the United States of America" (Bykov 1950, p. 768) were displayed at the congress by the fact that they sent 400 delegates, as compared with the USSR's 13.[27] The general gist of Bykov's report is that, whilst the conference's various attendees hung on the every word of the Soviet speakers, all other lectures (most particularly the American ones) were trivial and confused. There was, Bykov claimed, "not even a trace of that creative, sharp, critical attitude which marks free scientific discussions here" (Bykov 1950, p. 769). It should be noted that a report of the congress by a Westerner, published in the same issue of *Science* as Bykov's, is rather more measured in its tone (Gerard 1950).

Although, for the most part, the reports of one-off travellers to the USSR were more sober in tone than that of Bykov, they nevertheless had the slight drawbacks that they offered only a glimpse of scientific life in the Soviet Union, and that they could inadvertently be tinted by any propaganda that the traveller had been exposed to during their visit. The alternative to such impressionistic reports was of course the publication of detailed accounts of Soviet science, based on exhaustive research. We have seen already that such reports were not confined merely to the period of the Cold War, but had in fact appeared occasionally in earlier decades. These prior accounts, however, were perhaps motivated purely by curiosity, rather than fear, since the scientists of the West seem to have had only a limited esteem for Soviet science in the years before the Second World War. With regard to the war years, we have seen that surveys of Soviet research in particular fields did occasionally slip through the surrounding rhetoric of solidarity, particularly surveys on medical matters (Sect. 2.4). Indeed, it was probably during the Second World War that the importance of such surveys was first realised, or at least noted explicitly; we find, for example, the following remarks in a British review of an American wartime survey of Soviet science:

> In some way or other, British scientific workers must learn more about what their Russian colleagues are doing. Admittedly the difficulties are considerable. It is too much to expect British scientific workers to learn Russian on any extensive scale, and it is unfair to Russian science to judge the papers solely by the short summaries in English given at the ends. Perhaps the best way to ensure a proper appreciation of the work would be to arrange for the publication of systematic accounts of specific subjects, ... and of translations of specially important papers. (Russell 1944, p. 591)

(Footnote 26 continued)
Further examples of reports are provided by Aleksandrov (1977) and Gelfond (1977), although these were written decades after the trips that they describe.

[27] The figure of 13 for Soviet attendance comes from Gerard (1950).

The issue of Western scientists learning Russian will be taken up in Sect. 4.4, whilst the translation of Russian work is the subject of Sect. 4.5.

One way or another, the suggestion made by the above reviewer was gradually taken up in the decades following the Second World War: the brief appraisals of Soviet technological abilities that had been compiled following the launch of Sputnik were now joined in the literature by detailed surveys (in both book and article form) of Soviet academic science, specially commissioned by Western scientific organisations. Amongst the vast number of available such surveys, we find both accounts of the Soviet scientific establishment in general,[28] and discipline-specific reports, some of them by Soviet authors.[29] Moreover, symposia on the subject of Soviet science came to be held in the West, such as that organised by the American Association for the Advancement of Science in December 1951 (Christman 1952)[30]; as well as dealing with the specifics of Soviet science, this symposium also touched upon such issues as intellectual freedom in the USSR (Volin 1952). In the introduction to one subject-specific survey volume, we find the following comment, which sums up the purpose of any of these books:

> it is not certain that our general scientific community quite realizes the intense scientific activity that prevails in the Soviet Union. It is hoped that this report will do its share toward clearing the fog. (LaSalle and Lefschetz 1962, p. v)

I have yet to discover any comprehensive Soviet surveys of Western science. It is not unreasonable to speculate that such reports were indeed produced, but, like some of the reports of returning travellers, they may not have been intended for general consumption, which may explain why they are difficult to discover now.

Alongside established science and scientific culture, Western investigators also took an interest in how the USSR was preparing the next generation of scientists (indeed, citizens more generally). Thus, Soviet education received a great deal of attention in the West. Amongst the most comprehensive accounts of the Soviet education system, from primary to postgraduate level, with further details on employment prospects, are two reports prepared by Nicholas de Witt for the US National Science Foundation: *Soviet Professional Manpower: Its Education, Training and Supply* (1955) and *Education and Professional Employment in the USSR* (1961).

---

[28] See, for example, Leontief (1945), Oster (1949), Turkevich (1956), Rabinowitch (1958), Anon (1969a), and White (1971). Such accounts, which in many cases were written merely as technical guides to Soviet scientific organisation, stand alongside works of a more academic nature; we have, for example, Joravsky (1961), Lewis (1972), National Council (1975), Berry (1988), and Holloway (1994, 1999). See also the resources on Soviet science that have already been cited: those in note 2 on p. 1, together with Gerovitch (2002), Kojevnikov (2004), and Medvedev (1979).

[29] See, for example, Vinogradov (1947), Kline (1952), London (1954), Küng (1961), Spitsyn (1961), and Arkhimovich (1962).

[30] Recall from Sect. 2.4 that another (more propagandistic) conference on Soviet science had been held in London a few years earlier: see p. 21 and the references thereupon.

As with accounts of Soviet scientific activity, reports on education ranged from the general to the discipline-specific.[31]

In Sect. 2.5, we noted the launch in 1941 of the US journal *The Russian Review*, born of a wartime desire to learn more about the USSR, which carried (indeed, carries) articles on general Russian culture, which have often included accounts of Russian science. In fact, this is just one of several such journals to have appeared in the years of and following the Second World War. We also have, for example, *Slavic Review*, which began life in 1940 as the American series of the British *Slavonic Yearbook*, and is published by the American Association for Slavic, East European, and Eurasian Studies.[32] Another journal, *Studies on the Soviet Union*, was published in Munich by the Institute for the Study of the USSR from 1957 to 1971. All of these journals have carried articles of a scientific nature, some of which are cited in the present book. Between 1937 and 1992, non-technical (and often uncritical) outlines of Soviet scientific activity were also conveyed to British readers through *The Anglo-Soviet Journal*,[33] an organ of the SCR (see Sect. 2.2). The scientific articles published in *The Anglo-Soviet Journal* are comparable to those that appeared in *Britanskii soyuznik*, as discussed in Sect. 2.5.

Occupying the ground between abstracts and fuller surveys of Soviet scientific work, we have the digest *Russian Technical Literature*, which was published in Paris from 1960 by The Directorate of Scientific Affairs of the Organisation for Economic Cooperation and Development. This journal, which was styled "[a] bulletin aiming to create interest in the use of Russian and other Eastern European scientific and technical publications", carried short reports (with references to more comprehensive accounts) of Soviet scientific developments and organisation,[34] as well as announcements of exchange visits, new publications (including scientific language guides, of the types to be discussed in Sect. 4.4), Russian language courses, and new translation programmes, amongst other things. In 1964, *Russian Technical Literature* became *Science East to West* in light of its increasing treatment also of Chinese and Japanese scientific sources. The general ethos of *Russian Technical Literature/Science East to West* appears to have been to help Western scientists to access Soviet (later, Chinese and Japanese) scientific literature in any way possible (see, for example, Anon 1963), either in the original or in the rapidly proliferating translations that we will discuss in Sect. 4.5.

As remarked above, a key feature of *Russian Technical Literature*, and, indeed, of many of the surveys of Soviet science that have already been cited, is the fact

---

[31] See, for example, Anisimov (1950), Apanasewicz and Rosen (1964), Bernstein (1948), Kline (1957), Joravsky (1983), Litchfield et al. (1958), and Thwaites (1968). Further sources, published in the West, but written by Soviet authors, are Gnedenko (1957) and Petrovskii (1964).

[32] Vol. 1 appeared in 1940 as the *Slavonic Yearbook. American Series*. Vols. 2 and 3, of 1943 and 1944, were published under the name *Slavonic and East European Review. American Series*, which subsequently became the *American Slavic and East European Review*. The title *Slavic Review* was adopted from vol. 20, no. 3 (1961) onwards; see Byrnes (1976, p. 22).

[33] See Lygo (2013, p. 590). Scientific articles in the journal include Betenov (1946), Vavilov (1947), Morton (1948), Bernal (1950), Ambartsumyan (1955), and Anon (1969b).

[34] See, for example, Kowalewski (1963) and Anon (1964a, b, c).

that their authors did not seek merely to inform their readers about Soviet science, but also to equip them with the resources to be able to investigate further for themselves. Beyond the subject-specific surveys, guides to Soviet scientific literature were produced, ranging from simple guides to library holdings, such as those noted in Sect. 3.1, to articles outlining strategies for searching the Soviet literature. Thus, for example, the volume *Recent Soviet Contributions to Mathematics* features a guide to Soviet mathematical journals (Steeves 1962). Indeed, there are many other such articles, across a range of subject areas.[35] As with technical surveys of Soviet science, we again find specially-commissioned reports to aid Western researchers,[36] and we even encounter descriptions of the library services available in the USSR (for example, Horecky 1959, 1962)—Western researchers could thus not only learn of Soviet scientific advances, but also gain some idea of what resources were available to their Soviet counterparts. All such guides remain extremely useful in the historical study of Soviet science.[37] Similar such guides, this time on the availability of translations of Russian papers, will appear in Sect. 4.5.

References

Abelson, P.H.: International meetings. Science **154**(3747), 21 Oct, 341 (1966)

Adams, S., Rogers, F.B. (eds.): Guide to Russian Medical Literature. US Department of Health, Education, and Welfare, Washington, DC (1958)

Aleksandrov, P.S.: Memories of Göttingen. Istor.mat. issled., no. 22, 242–245 (1977) (in Russian)

Ambartsumyan, V.A.: The problem of the origin of stars. Anglo-Soviet J., Sept, 7–16 (1955)

Anisimov, O.: The Soviet system of education. Russian Rev. **9**(2), 87–97 (1950)

Anon: Periodicals received by the Moscow Mathematical Society. Mat. sb. **21**(4), 779–781 (1901a) (in Russian)

Anon: List of individuals and institutions to whom were sent volume XXI of 'Matematicheskii sbornik'. Mat. sb. **21**(4), 782–785 (1901b) (in Russian)

Anon: Scientific publications for Russia. Nature **107**(2697), 7 Jul, 594–594 (1921)

Anon: Soviet medical and scientific men. Nature **151**(3833), 17 Apr, 444 (1943)

Anon: British doctors in Russia: experiences and impressions. Brit. Med. J. **2**(4892), 9 Oct, 863–865 (1954)

Anon: Foreign Technical Information Center. Science **127**(3294), 14 Feb, 332–333 (1958)

Anon: Survey of Soviet science literature. Science **130**(3371), 7 Aug, 324 (1959a)

Anon: The Report of the United States Public Health Mission to the Union of Soviet Socialist Republics: Including Impressions of Medicine and Public Health in Several Soviet Republics, August 13 to September 14, 1957. US Department of Health, Education, and Welfare, Washington, DC (1959b)

Anon: Biochemistry in Russia: impressions gained at the Fifth International Congress of Biochemistry in Moscow. Brit. Med. J. **2**(5253), 9 Sept, 701–703 (1961)

---

[35]See, for example, Adams and Rogers (1958), Barker (1954), Berry (1988, Chap. 12), Chard (1969), Hoseh (1961), Lieberman (1987), Mackay (1954), Stubbs (1957), and Zhavoronkov (1956).

[36]See, for example, Gorokhoff (1962) and Wiggins (1972).

[37]Indeed, we might add to these two further articles, written specifically for historians of Soviet science: Demidov (1993), and the 'Bibliographic Essay' in Graham (1993, pp. 293–306).

Anon: The Cunningham Amendment. ACLS Newsl. **13**(4), 13–16 (1962a)

Anon: Part I: a general appraisal of mathematics in the USSR. In: LaSalle and Lefschetz (1962), pp. 3–13 (1962b)

Anon: Decoding Russian titles and sentences. Russian Tech. Lit., no. 12, Oct, 25–27 (1963)

Anon: Research on corrosion in Russia. Science East to West, no. 14, Apr, 13–14 (1964a)

Anon: Technical education in the U.S.S.R. Science East to West, no. 15, Jul, 1–8 (1964b)

Anon: Scientific and technical information in the Soviet Union. Science East to West, no. 15, Jul, 9–11 (1964c)

Anon: Russian science policy. Nature **221**(5178), 25 Jan, 307–308 (1969a)

Anon: The Ukrainian Academy of Sciences. Anglo-Soviet J., Sept, 6–8 (1969)

Anon: On the occasion of the centenary of the birth of Vladimir Ilich Lenin. Uspekhi mat. nauk **70**(2), 3–4 (1970) (in Russian); English trans.: Russian Math. Surveys **25**(2), 1–2 (1970)

Anon: Soviet science not quoted. Nature **299**(5884), 14 Oct, 568 (1982)

Apanasewicz, N., Rosen, S.M.: Soviet Education: A Bibliography of English-language Materials. Report no. BULL-1964-29; 0E-14101. Office of Education (DHEW), Washington, DC (1964)

Arkhimovich, A.: A survey of Soviet agriculture. Studies on the Soviet Union **2**(3), 51–63 (1962)

Armstrong, J.A.: The Soviet intellectuals: observations from two journeys. Stud. Soviet Union **1**(1), 25–37 (1961)

Arutjunov, N.B.: The system of international exchange of scientific and technological information: the USSR's participation. UNESCO J. Inform. Sci. Librarianship Archives Admin. **1**(3), 210–214 (1979)

Ashby, E.: Scientist in Russia. Penguin Books, Harmondsworth (1947)

Baker, D.B., Horiszny, J.W., Metanomski, W.V.: History of abstracting at Chemical Abstracts Service. J. Chemical Inform. Comp. Sci. **20**(4), 193–201 (1980)

Barker, G.R.: Sources of Russian economic information. Aslib Proc. **6**(2), 101–110 (1954)

Bateson, W.: Science in Russia. Nature **116**(2923), 7 Nov, 681–683 (1925)

Benedict, F.G.: Russian research in metabolism. Science **29**(740), 5 Mar, 394–395 (1909)

Benjamin, L.T., Jr., VandenBos, G.R.: The window on psychology's literature: a history of *Psychological Abstracts*. Amer. Psychol. **61**(9), 941–954 (2006)

Bernal, J.D.: Science in the USSR to-day. Anglo-Soviet J., Sept, 4–17 (1950)

Bernstein, M.: Higher education in the USSR during and after the war. Educational Forum **12**(2), 209–212 (1948)

Berry, M.J. (ed.): Science and Technology in the USSR. Longman (1988)

Betenov, C.: Azerbaidjan Academy of Sciences. Anglo-Soviet J., Jun, 15–18 (1946)

Beyer, R.T.: Hurdling the language barrier. Physics Today **18**(1), 46–52 (1965)

Beyerly, E.: A Russian abstracting service in the field of sciences: *Referativnyĭ zhurnal*. Aslib Proc. **8**(1), 135–140 (1956)

Bloom, H.: The end of samizdat? The Soviet Union signs the Universal Copyright Convention. Index on Censorship **2**(2), 3–18 (1973)

Bockris, J.O'M.: A scientist's impressions of Russian research. The Reporter **18**(14), 20 Feb, 15–17 (1958)

Boffey, P.M.: Censorship angers scientists: Pentagon security move bars 100 technological papers. NY Times, 5 Sept, 1, 16 (1982)

Bykov, C.: Soviet physiologists at the International Congress. Izvestiya, 23 Sept (1950) (in Russian); English trans.: Science **112**(2921), 22 Dec, 768–769 (1950)

Byrnes, R.F.: Soviet-American Academic Exchanges, 1958–1975. Indiana Univ. Press (1976)

Cadbury, D.: Space Race: The Untold Story of Two Rivals and Their Struggle for the Moon. Fourth Estate, London (2005)

Carey, W.D.: Censorship, Soviet style. Science 219(4587), 25 Feb, 911 (1983)

Chard, C.C: Archeology in the Soviet Union. Science **163**(3869), 21 Feb, 774–779 (1969)

Charlier, R.H., Dietz, R.S.: Oceanography: two reports on the recent International Congress in Moscow. Science **153**(3742), 16 Sept, 1421–1428 (1966)

Chebukov, D.E., Izaak, A.D., Misyurina, O.G., Pupyrev, Yu.A., Zhizhchenko, A.B.: Math-Net.Ru as a digital archive of the Russian mathematical knowledge from the XIX century to today. In: Carette, J., Aspinall, D., Lange, C., Sojka, P., Windsteiger, W. (eds.) Intelligent Computer Mathematics, pp. 344–348. Springer (2013)

Christman, R.C. (ed.): Soviet Science: A Symposium Presented on December 27, 1951, at the Philadelphia Meeting of the American Association for the Advancement of Science. Amer. Assoc. Adv. Sci. (1952)

Danckwerts, P.: To Russia with science. New Scientist 100(1389–1390), 22–29 Dec, 943 (1983); ibid. 101(1391), 5 Jan, 39 (1984)

David-Fox, M.: Showcasing the Great Experiment: Cultural Diplomacy and Western Visitors to the Soviet Union, 1921–1941. Oxford Univ. Press (2012)

Demidov, S.S.: A brief survey of literature on the development of mathematics in the USSR. In: Zdravkovska, S., Duren, P.L. (eds.) Golden Years of Moscow Mathematics, pp. 245–262. Amer. Math. Soc./London Math. Soc. (1993)

Demidov, S.S.: 'Matematicheskii sbornik' in 1866–1935. Istor.-mat. issled., no. 1(36), 127–145 (1996) (in Russian)

DuS., G.: Scientific information in the U.S.S.R. Science 124(3223), 5 Oct, 609 (1956)

Fernberger, S.W.: Publications, politics and economics. Psychol. Bull. 35(2), 84–90 (1938)

Forman, P.: Scientific internationalism and the Weimar physicists: the ideology and its manipulation in Germany after World War I. Isis 64(2), 150–180 (1973)

Frankel, M.: Soviet amasses scientific data; special institute prepares digests of technical publications of world. NY Times, 15 May, 9 (1958)

Friedman, M.D.: On procuring Russian literature. Science 155(3761), 27 Jan, 400 (1967)

Furaev, V.K.: Soviet-American scientific and cultural relations (1924–1933). Voprosy istorii, no. 3, 41–57 (1974) (in Russian); English trans.: Soviet Stud. Hist. 14(3), 46–75 (1975–1976)

Gak, A.M.: V. I. Lenin and the development of the international cultural and scientific ties of Soviet Russia in 1920–1924. Voprosy istorii, no. 4, 196–204 (1963) (in Russian)

Garfield, E.: Some implications of the Soviet Union's becoming party to the Universal Copyright Convention. Current Contents, no. 15, 11 Apr, 5–7 (1973); (also in: Essays of an Information Scientist, vol. 1, pp. 428–430. ISI Press, Philadelphia (1962–73))

Gelfond, A.O.: Some impressions of a scientific trip to Germany in 1930. Istor.-mat. issled., no. 22, 246–251 (1977) (in Russian)

Gerard, R.W.: The Eighteenth International Physiological Congress. Science 112(2921), 22 Dec, 767 (1950)

Gerovitch, S.: From Newpeak to Cyberspeak: A History of Soviet Cybernetics. MIT Press (2002)

Gnedenko, B.V.: Mathematical education in the U.S.S.R. Amer. Math. Monthly 64(6), 389–408 (1957)

Gnedenko, B.V.: V. I. Lenin and methodological questions of mathematics. Uspekhi mat. nauk 70(2), 5–14 (1970) in Russian; English trans. Russian Math. Surveys 25(2), 3–12 (1970)

Gordin, M.D.: Scientific Babel: The Language of Science from the Fall of Latin to the Rise of English. Profile Books (2015)

Gorokhoff, B.I.: Providing U.S. Scientists with Soviet Scientific Information. National Science Foundation Washington DC (1959); revised ed. (1962)

Graham, L.R.: Science in Russia and the Soviet Union: A Short History. Cambridge Univ. Press (1993)

Greenberg, J.: 'Remote censoring': DOD blocks symposium papers. Science News 122(10), 4 Sept, 148–149 (1982)

Gregory, R.A., Wright, C.H.: Scientific literature for Russia. Nature 109(2729), 16 Feb, 208 (1922)

Hamblin, J.D.: Science in isolation: American marine geophysics research, 1950–1968. Physics in Perspective 2(3), 293–312 (2000)

Hollings, C.: Mathematics across the Iron Curtain: A History of the Algebraic Theory of Semigroups. Amer. Math. Soc., Providence, Rhode Island (2014)

Holloway, D.: Stalin and the Bomb: The Soviet Union and Atomic Energy, 1939–1956. Yale Univ. Press (1994)

Holloway, D.: Physics, the state, and civil society in the Soviet Union. Hist. Stud. Phys. Biol. Sci. **30**(1), 173–192 (1999)

Horecky, P.L.: Libraries and Bibliographic Centers in the Soviet Union. Indiana Univ. Publ., Bloomington IN (1959)

Horecky, P.L.: Soviet library literature: a general survey of some recent trends and examples. Library J. **87**(4), 715–719 (1962)

Horecky, P.L.: The Slavic and East European resources and facilities of the Library of Congress. Slavic Rev. **23**(2), 309–327 (1964)

Hoseh, M.: Scientific and technical literature of the USSR. In: Gould, R.F. (ed.) Searching the Chemical Literature, pp. 144–171. American Chemical Society, Washington DC (1961)

Huxley, J.: A Scientist among the Soviets. Chatto and Windus, London (1932)

Joravsky, D.: Soviet Marxism and Natural Science, 1917–1932. Routledge and Kegan Paul, London (1961)

Joravsky, D.: The Stalinist mentality and the higher learning. Slavic Rev. **42**(4), 575–600 (1983)

Josephson, P.R.: Soviet scientists and the state: politics, ideology, and fundamental research from Stalin to Gorbachev. Social Research **59**(3), 589–614 (1992)

Kameneva, O.D.: Cultural rapprochement: the U.S.S.R. Society for Cultural Relations with Foreign Countries. Pacific Affairs **1**(5), 6–8 (1928)

Kellogg, V.: Russian scientific literature. Science **56**(1437), 14 Jul, 45 (1922)

Kline, J.R.: Soviet Mathematics. In: Christman (1952), pp. 80–84

Kline, G.L. (ed.): Soviet Education. Routledge and Kegan Paul, London (1957)

Kojevnikov, A.B.: Stalin's Great Science: The Times and Adventures of Soviet Physicists. Imperial College Press (2004)

Kolata, G.: Export control threat disrupts meeting. Science **217**(4566), 24 Sept, 1233–1234 (1982)

Kowalewski, J.: Low-grade ore beneficiation in the Soviet Union. Russian Tech. Lit., no. 11, Jul, 1–13 (1963)

Kraus, D.H.: The Slavic and Central European Division and the Slavic Collections of the Library of Congress. Federal Linguist **7**(1–4), 4–11 (1976)

Kraus, D.H.: The Slavic Collections of the U.S. Library of Congress. UNESCO J. Inform. Sci. Librarianship Archives Admin. **1**(3), 184–190 (1979)

Krementsov, N.: International Science between the World Wars: The Case of Genetics. Routledge, New York and London (2005)

Kuhterina, P.P.: Reference services in the All-Union State Library of Foreign Literature, USSR. UNESCO J. Inform. Sci. Librarianship Archives Adm. **2**(1), 42–49 (1980)

Küng, G.: Mathematical logic in the Soviet Union (1917–1947 and 1947–1957). Studies in Soviet Thought **1**, 39–43 (1961)

Lapko, A.F., Lyusternik, L.A.: Lenin, science and education. Uspekhi mat. nauk **70**(2), 15–60 (1970) (in Russian); English trans.: Russian Math. Surveys **25**(2), 13–76 (1970)

LaSalle, P., Lefschetz, S. (eds.): Recent Soviet Contributions to Mathematics. Macmillan (1962)

Lehmer, D.H.: A half century of reviewing. In: Duren, P.L. (ed.) A Century of Mathematics in America, vol. 1, pp. 265–266. Amer. Math. Soc. (1989)

Leontief, W.W., Sr.: Scientific and technological research in Russia. Amer. Slavic East Europ. Rev. **4**(3/4), 70–79 (1945)

Levich, Y.: Trying to keep in touch. Nature **263**(5576), 30 Sept, 366 (1976)

Lewis, R.A.: Some aspects of the research and development effort of the Soviet Union, 1924–35. Social Studies of Science **2**, 153–179 (1972)

Lieberman, E.R.: Where to look for Soviet MS/OR articles: a guide to English language sources and abstracts. Interfaces **17**(4), 85–89 (1987)

Litchfield, E.H., Mettger, H.P., Gideonse, H.D., Glennan, T.K., Harnwell, G.P., Malott, D.W., Murphy, F.D., Scaife, A.M., Sparks, F.H., Wells, H.B.: Report on Higher Education in the Soviet Union. Univ. Pittsburgh Press (1958)

Lohwater, A.J.: Mathematics in the Soviet Union. Science **125**(3255), 17 May, 974–978 (1957)

London, I.D.: Research on sensory interaction in the Soviet Union. Psychol. Bull. **51**(6), 531–568 (1954)

Lygo, E.: Promoting Soviet culture in Britain: the history of the Society for Cultural Relations between the Peoples of the British Commonwealth and the USSR, 1924–1945. Modern Lang. Rev. 108(2), 571–596 (2013)

Mackay, A.L.: Sources of Russian scientific information. Aslib Proc. **6**(2), 101–110 (1954)

Medvedev, Zh.A.: The Medvedev Papers: The Plight of Soviet Science Today. Macmillan, London (1971)

Medvedev, Zh.A.: Soviet Science. Oxford Univ. Press (1979)

Menschutkin, B.: Co-operation in Russian and British scientific undertakings. Nature **99**(2478), 26 Apr, 168–169 (1917)

Montagu of Beaulieu, Barker, E., Cathcart, E.P., Eddington, A.S., Gollancz, I., Gregory, R.A., Mitchell, P.C., Pares, B., Schuster, A., Sherrington, C.S., Shipley, A.E., Wells, H.G., Woodward, A.S., Wright, C.H.: The British Committee for Aiding Men of Letters and Science in Russia. Nature **106**(2671), 6 Jan, 598–599 (1921)

Morton, A.G.: Biology in the Soviet Union. Anglo-Soviet J., Dec, 5–8 (1948)

National Council of American-Soviet Friendship: Science in Soviet Russia. Arno Press (1975)

Oster, G.: Scientific research in the U.S.S.R: organization and planning. Ann. Amer. Acad. Political Social Sci. **263**, 134–139 (1949)

Panel on Scientific Communication and National Security, Committee on Science, Engineering, and Public Policy, Institute of Medicine, Policy and Global Affairs, National Academy of Sciences, National Academy of Engineering: Scientific Communication and National Security. The National Academies Press, Washington, DC (1982)

Panov, D.: Scientific abstracting in the U.S.S.R. Science **121**(3148), 29 Apr, 627–628 (1955)

Panov, D.: Institute of Scientific Information of the Academy of Sciences of the U.S.S.R. J. Documentation **12**(2), 94–100 (1956)

Park, R.L.: The muzzling of American science. Academe **72**(5), 19–23 (1986)

Penfield, W.: A glimpse of neurophysiology in the Soviet Union. Canad. Med. Assoc. J. **73**, 891–899 (1955)

Peterson, I.: Silencing science for security. Science News **121**(3), 16 Jan, 35 (1982a)

Peterson, I: Hearing considers impact of secrecy proposals on science. Science News **121**(14), 3 Apr, 229 (1982b)

Peterson, I.: A question of scientific free speech. Science News **122**(25/26), 18–25 Dec, 396–397 (1982c)

Petrovskii, I.G.: Higher education in Moscow. Internat. Sci. Tech., no. 29, 33–37 (1964)

Piaget, J.: Some impressions of a visit to Soviet psychologists. Acta Psychol. **12**, 216–230 (1956)

Pitcher, E.: A history of the second fifty years. In: American Mathematical Society, 1939–1988, vol. I, pp. 69–89. Amer. Math. Soc. (1988)

Rabinowitch, E.: Soviet science — a survey. Problems and Communism **7**(2), 1–9 (1958)

Razran, G.S.: Offprints for the scientific men of Soviet Russia. Science **96**(2488), 4 Sept, 231 (1942)

Rich, V.: Russia's curtained window on the West. Nature **249**(5457), 7 Jun, 502–504 (1974)

Rich, V.: Soviet meetings, great and small. Nature **256**(5514), 17 Jul, 160–161 (1975)

Richert, N.: Mathematical Reviews celebrates 75 Years. Notices Amer. Math. Soc. **61**(11), 1355–1356 (2014)

Richmond, Y.: Cultural Exchange and the Cold War: Raising the Iron Curtain. Pennsylvia State Univ. Press (2003)

Russell, E.J.: Science in the U.S.S.R.: an American survey. Nature **154**(3915), 11 Nov, 590–591 (1944)

Schuster, L.F.: Literature for men of letters and science in Russia. Nature **106**(2675), 3 Feb, 728 (1921)

Schwartz, H.: A Soviet 'curtain' hung over science. NY Times, 9 Oct (1951)

Serpinsky, V.V.: Referativny zhurnal *Khimiya* (Abstracts Journal *Chemistry*) of the Academy of Sciences of the U.S.S.R. J. Documentation **12**(2), 100–106 (1956)

Shapley, D.: Pentagon blocks open exchange: Weinberger in optics meeting censorship. Nature **299**(5881), 23 Sept, 289–290 (1982)

Sherrod, J.: The Library of Congress. Science **127**(3304), 25 Apr, 958–959 (1958)

Siegmund-Schultze, R.: "Scientific control" in mathematical reviewing and German–U.S.-American relations between the two world wars. Historia Math. **21**, 306–329 (1994)

Siegmund-Schultze, R.: One hundred years after the Great War (1914–2014): a century of breakdowns, resumptions and fundamental changes in international mathematical communication. In: Jang, S.Y., Kim, Y.R., Lee, D.-W., Yie, I. (eds.) Proceedings of the International Congress of Mathematicians Seoul 2014, vol. IV, pp. 1231–1253. Gyeong Munsa (2014)

Sinclair, G.W.: A report on Biological Abstracts. J. Paleontology **27**(2), 297 (1953)

Solomon, S.G., Krementsov, N.: Giving and taking across borders: the Rockefeller Foundation and Russia, 1919–1928. Minerva **39**, 265–298 (2001)

Spitsyn, V.I.: Soviet Chemistry Today: Its Present State and Outlook for the Future. National Academy of Sciences-National Research Council, Washington DC (1961)

Steeves, H.A.: Part X: Russian journals of mathematics. In: LaSalle and Lefschetz (1962), pp. 303–315 (1962)

Stubbs, A.E.: The dissemination of the knowledge of Soviet scientific work in Western countries. Aslib Proc. **9**(11), 333–340 (1957)

Tareev, B.M.: Methods of disseminating scientific information, and science information activities in the USSR. Amer. Documentation **13**(3), 338–343 (1962)

Thwaites, B.: Mathematical education in Russian schools: a report on a visit to Russia in August 1966. Math. Gaz. **52**(382), 319–327 (1968)

Turkevich, J.: Soviet science in the post-Stalin era. Ann. Amer. Acad. Political Social Sci. **303**, 139–151 (1956)

Vavilov [*sic*]: Soviet science in the new five-year plan. Anglo-Soviet J., Dec, 5–11, 45–8 (1947)

Vinogradov, I.: Soviet mathematicians. Synthese **5**(11/12), 501–503 (1947)

Vlachý, J.: Publication output of Soviet physics. Czechoslovak J. Phys. B **29**(3), 357–360 (1979)

Volin, L.: Science and intellectual freedom in Russia. In: Christman (1952), pp. 85–99

Walsh, W.B.: Some judgments on Soviet science. Russian Rev. **19**(3), 277–285 (1960)

White, S. (ed.): Guide to Science and Technology in the USSR: A Reference Guide to Science and Technology in the Soviet Union. Francis Hodgson (1971)

Wiggins, G.: English-language Sources for Reference Questions Related to Soviet Science (With an Emphasis on Chemistry). Univ. Illinois Grad. School of Library Science Occasional Papers, no. 102, Jun (1972)

Zhavoronkov, N.M.: Publication of literature on applied chemistry in the Soviet Union. J. Documentation **12**(2), 106–113 (1956)

Chapter 4
Linguistic Access to Publications

Abstract We turn in this chapter to linguistic matters, which we set in the context of the infamous 'foreign-language barrier'. The specific issues considered here are the use of foreign languages, and the appearance of foreign authors, in Soviet journals, Russian-language ability amongst Western scientists, and the translation of scientific works.

Keywords Foreign-language barrier · Soviet use of foreign languages · Western use of Russian · Soviet science journals · Scientific translation

The general impression that the reader ought to have gained from Chap. 3 is that physical access to published sources from the opposite side of the Iron Curtain, though by no means comprehensive, was, generally speaking, not a major problem. A more significant problem, at least in some quarters, was that of *linguistic* accessibility. In this chapter, I argue that whilst *physical* accessibility of Western materials might occasionally have been a problem for Soviet scientists (thanks to such issues as censorship), it was *linguistic* access that was the greater problem for Western scientists when it came to the handling of Soviet sources.

4.1 The Foreign-Language Barrier

I wish, first of all, to place the following discussions of language into the context of the infamous 'foreign-language barrier': the fact that the multiplicity of languages used in international scientific discourse means that a sizeable proportion of the world's scientific literature remains inaccessible to a large number of the world's scientists.[1] A great deal has been written about the foreign-language barrier, particularly in the second half of the 20th century, when Cold War rivalries, and the

---

[1] There is of course a broader foreign-language barrier, beyond science, but I confine my attention to the scientific context. I also approach the problem largely from a native-English-speaking perspective. A major recent publication in this area is Gordin (2015).

C.D. Hollings, *Scientific Communication Across the Iron Curtain*,
SpringerBriefs in History of Science and Technology,
DOI 10.1007/978-3-319-25346-6_4

enormous growth in scientific publishing,[2] appear to have high-lighted the need to access foreign scientific literature.[3] Indeed, concerns were grave enough that we find, amongst the various materials on this subject, reports specially commissioned by such bodies as UNESCO and the ICSU.[4] Naturally, Russian scientific literature received a great deal of attention, although this was not uniform across all disciplines. Indeed, the *Russian-language* barrier was noted as early as 1909 by the American physiologist Francis Gano Benedict,[5] who lamented that significant Russian work on metabolism was going unread elsewhere in the world, except perhaps for short accounts in French and German abstracting journals (Benedict 1909). We will note some of Benedict's efforts to remedy this situation in Sect. 4.5. In later decades, the enormous size of the Soviet scientific corpus[6] was such that, even before analysis, abstracting, or translation could take place, great efforts were required simply to keep track of the available material (see, for example, Zikeev 1963).

The various sources dealing with language problems, covering several decades and a range of disciplines, make for rather dispiriting reading, for they say much the same thing over and over again: the foreign-language barrier is identified,[7] an argument is made for the need to overcome it, current measures to do so within the country and/or discipline of interest are surveyed, and recommendations are made for further improvement. But when one picks up a later treatment of the foreign-language barrier, written, say, a decade later, one finds that all the same problems remain: the recommendations of the previous report have not been implemented (at least not comprehensively), and there is little for the new author to do but to make them afresh. Little appears to change.

Various aspects of the above-mentioned reports on the foreign-language barrier will emerge over the course of this chapter. It is nevertheless useful to record some of their main points here. To begin with, it was observed that the relevance, or at least the perceived relevance, of the scientific output of different nations varied from discipline to discipline. Thus, for example, one survey found that, on the whole, British scientists made very little use of foreign literature, with the possible exception of that in French (Anderson 1978). Other figures indicate that Western biologists appear to have relied most heavily on materials in English, and, indeed,

---

[2]See Bourne (1962), Barr (1967), and Tschirgi (1973). Indeed, some commentators have cited information overload as a more serious problem than the language barrier: Garfield (1983).

[3]See, for example, Hanson (1962), Holmstrom (1962), Wood (1967), Hunter (1970), Hutchins et al. (1971a, b), Kertesz (1974), Chan (1976), and Large (1983). Indeed, the foreign-language barrier in scientific communication remains a matter of current concern: see, for example, Ammon (2006), Montgomery (2013), and Sloan and Alper (2014, Chap. 4).

[4]See, respectively, UNESCO (1957) and Anon (1962).

[5]See note 8 on p. 10.

[6]See, for example, the comments in Gordin (2015, p. 217).

[7]Often with the use of a statement that makes one pause and scratch one's head, such as: "at least 50 % of scientific literature is in languages which more than half the world's scientists cannot read" (UNESCO 1957, p. 13).

were doubtful as to the worth of the Russian literature.[8] The value of Soviet medical and social science research was also questioned, with the suggestion that such material was tainted by state ideology[9]; Soviet chemistry was sometimes treated likewise (Rathmann 1958). Indeed, we find similar such views throughout the material on the foreign-language barrier, together with pleas to give Soviet science a chance.[10] However, these negative attitudes often owed more to simple chauvinism than to objective assessment: take, for example, the West's complacency with regard to Soviet technical expertise "until the bleeps of Sputnik I sent Western scientists running to their Russian primers" (Anon 1977). It should be noted that such attitudes were not confined to the Soviet context: according to a study published in the early 1980s, Dutch biochemists were of the opinion that there was nothing worth reading in the French literature (Gordon and Santman 1981, p. 186).

The use of foreign literature by scientists in various fields has also been the subject of much study, with some commentators observing a disparity between the professed language skills of scientists and their inclination to use them to access foreign literature (Chan 1976, pp. 318–319): in many cases, the scientists felt that there was already too much material for them to assimilate in their native language, let alone in any others. Foreign-language material held by scientific libraries was thus felt to be little-used, and hence poorly cited. This therefore led to a language bias in scientific citations that did not necessarily reflect the true status of the literature—a study of the mid-1950s, for example, asserted that the citation of the respective English- and German-language materials by Western psychologists, chemists and physicists was out of proportion to those languages' representation in the available literature (Louttit 1955, 1957). Somewhat later, a citation analysis of 1990 revealed that 98.8 % of citations of Russian-language scientific papers were by Soviet authors (Garfield and Welljams-Dorof 1990, p. 14)—Westerners simply were not citing Soviet research, even after several decades of agonising over the foreign-language barrier. Again, judgements as to the value of the literature probably had a role to play here, alongside considerations of scientists' linguistic abilities.

As noted above, treatments of the foreign-language barrier typically include suggestions on how it might be overcome. First amongst these was usually the recommendation that the foreign-language skills of scientists be improved, either by the laying on of courses for established academics, or by strengthening the foreign-language requirements at, say, the postgraduate level (UNESCO 1957, Sect. 5.3). The question of Russian-language ability amongst Western scientists, and efforts to improve it, will be treated in Sect. 4.4. Connected with questions of improved language ability were the frequent calls for scientists to adopt a single global language,[11] whether it be an existing one (English, French, German, Spanish, and

---

[8] See Chan (1976, pp. 317, 319) or Large (1983, p. 35).

[9] See Hutchins et al. (1971a, p. 6) or Herner (1958).

[10] See, for example, Walsh (1960) or London (1957).

[11] See, for example, Couturat et al. (1910); a recent discussion of this issue may be found in Gordin (2015, Chaps. 4 and 5).

Latin, amongst others, all had their claims pressed)[12] or an artificial one (Esperanto being the front-runner: see Large 1985). However, such suggestions were usually not entirely practical, and appear to have been taken seriously only by a minority. For better or for worse, the 20th century saw English assume the mantle of *de facto* scientific *lingua franca* (Ammon 2006; Montgomery 2013). The suggestion that Russian should be adopted as an international auxiliary language does not appear to have been taken seriously outside the Soviet sphere of influence (Large 1985, p. 196), although it was used quite successfully in that capacity within the Soviet bloc (see the comments in Sect. 4.4), and was also adopted elsewhere in more restricted roles: as one of the official languages (along with English and French) of the International Mathematical Union, for example (Lehto 1998, p. 109). On the whole, however, Soviet nationalistic insistence upon the sole use of Russian in scientific papers (see Sect. 4.2) contributed to the sometime-isolation of Soviet science (Garfield and Welljams-Dorof 1990), particularly in light of the fact that, in addition, Russian has only rarely been used as a major language for international congresses.[13] In response to Soviet linguistic attitudes, those in the West were forced to adopt the measures outlined in Sects. 4.4 and 4.5.

In recognition of the fact that scientists (particularly native English-speaking scientists) had neither the time nor the inclination to improve their language skills, recommendations were also made regarding the publication of English-language abstracts of scientific papers in other languages. As we saw in Sect. 3.4, this practice was already widespread from an early date. Nevertheless, it was felt that such abstracts might yet be made more comprehensive, and made to cover a broader range of materials. A Western scholar would thus be able to decide whether a given foreign paper was relevant to his or her research without having to commission a full translation. Although (as we will see) the translation of foreign scientific papers was (and, in some cases, remains) a very popular and widespread solution to the foreign-language barrier, it appears to have been regarded only as a last resort by many authors of reports on foreign-language issues. Certainly, it is not an activity without its problems. Broadly speaking, two options were/are available: the translation only of selected papers, or the cover-to-cover translation of entire journals. The former requires the value judgement of a specialist, concerning which papers to translate, but the specialist will likely only speak for a section of the relevant community. The latter requires no such judgement, but is considerably more expensive, and will almost certainly result in the production of papers that will never be consulted in translation. In either case, the delays involved in the production of scientific translations were often deemed unsatisfactory. Thus, although one writer hailed (cover-to-cover) translations as a means to "dispel ignorance, overcome prejudice, and increase interest in foreign science" (Hanson 1962, p. 56), there has nevertheless been much debate over the years as to the cost-effectiveness and intellectual value of producing translations of scientific materials.

---

[12]See, for example, Castro (1975), Jaramillo (1975), and Ammon (1998); see also Large (1983, Chap. 8).

[13]See, for example, the comments in Medvedev (1971, p. 132).

4.2 Foreign Languages in Soviet Journals

The foreign-language barrier, as described in the preceding section, was in fact less of a problem in the earlier part of our period of interest, principally because Russian had yet to become the major academic language that nationalistic considerations would later make it.[14] Russian was certainly used in many periodicals: in the early years of the 20th century, it was necessary to write up in Russian any research that had been funded by the Russian state (Neswald 2013, p. 32). We also find conference proceedings in Russian: both Russian translations of lectures given by foreign delegates at conferences in Russia (see, for example, Congrès 1910), and also Russian versions of the proceedings of conferences held elsewhere (Anon 1910). Nevertheless, in the academic setting, Russian existed alongside other languages; for example, one Russian metabolism journal of the early 20th century was published in two versions: one Russian, one French (Benedict 1909). The *Bulletin* of the Moscow Society of Naturalists was published in French also. Indeed, from its inception in 1894, the major organ of the Academy of Sciences, *Извѣстія Императорской Академіи Наукъ* (also commonly known by the direct French translation of this Russian title: *Bulletin de l'Académie Impériale des sciences*) regularly carried articles not only in Russian, but also French, German, and (occasionally) English. Somewhat later, the Academy published its *Journal of Physics* entirely in English.[15] To turn our attention to authors, we note that the late-19th/early-20th century Russian scientists who were mentioned in passing in Sect. 2.1 all customarily published work in languages other than Russian (sometimes in Russia, sometimes not: see, for example, Lapo 2001, p. 47). The use of foreign languages in work published abroad is hardly surprising, so we confine our attention to domestic publication. We should note also that, perhaps as a result of the extensive Soviet abstracting activities discussed in Sect. 3.4, papers by Soviet authors generally contain many references to work in foreign languages—so much so that Western authors sometimes commented upon the ease with which Soviet researchers handled foreign sources (Tolpin et al. 1951). I will say more about language ability amongst Soviet scientists in Sect. 4.4.

The trend amongst Russian scientists of publishing their work in foreign languages continued well into the Soviet era. In order to illustrate this, let us once again take the journal *Matematicheskii sbornik*, founded by the Moscow Mathematical Society in 1866, and subsequently co-published with the Academy of Sciences (Demidov 1996; Lyusternik 1946). Since it was created as a forum for the Moscow Mathematical Society, *Matematicheskii sbornik* was a largely domestic, indeed *local*, concern in the early decades of its publication, and so, up to and including vol. 30 (1916–1918) of its original series, was produced almost entirely

---

[14]Although the push towards the exclusive use of Russian was mainly a product of Stalin's era, we perhaps see it foreshadowed in the foundation of many Russian-language scientific periodicals in the nationalistic atmosphere of the First World War (Kojevnikov 2002, p. 240).

[15]See Mackay (1954, pp. 102, 109) or Gordin (2015, p. 225). Other examples of Soviet journals produced in Western languages are noted in Kryuchkova (2001, pp. 410–411).

in Russian. It has been suggested that this situation was maintained, at least in part, by the nationalist sentiments of the journal's early-20th century editor N.V. Bugaev (1837–1903) (Svetlikova 2013, p. 24). This language policy began to change, however, from vol. 31 (1922–1924), under the editorship of D.F. Egorov (1869–1931). An international audience was now sought for the journal, not to mention an international standing; the editorial mentioned in Sect. 2.3 put this rather stridently some years later: "Soviet mathematics can and should have a journal of international significance".[16] Thus, as with the Academy journal mentioned above, other languages were admitted: vol. 31 featured twelve articles in French and one in English. Indeed, this trend became even stronger during the 1920s, and remained strong into the 1930s, as we can see from Fig. 4.1[17]: we observe, for example, that non-Russian papers were in the majority in vols. 33 (1926), 36 (1929), 37 (1930), and 38 (1931) (just). Note that I have omitted vol. 43 (1936) from Fig. 4.1 since the data for this year are skewed somewhat by the journal's publication of papers (many by foreign authors) from an international conference on topology (see Sect. 2.3). Observe further that non-Russian papers were in the majority just once more, in vol. 51 (1941), probably as a result of the USSR's entry into the Second World War, and the resultant feelings of 'scientific solidarity' discussed in Sect. 2.4; unsurprisingly, the use of German in the journal ceased after this volume. As Fig. 4.1 shows, the use of foreign languages continued throughout the 1930s and 1940s, although it began to slow: the final foreign-language paper appeared in vol. 63 (1947), after which the journal reverted exclusively to Russian. It should be noted that those authors who were writing for *Matematicheskii sbornik* in foreign languages included a number of foreign authors—I will return to this point in the next section.

In spite of its new policy of publishing papers in foreign languages, Russian remained at the core of *Matematicheskii sbornik*: many papers still appeared in Russian, and those papers that were published in foreign languages nevertheless carried a Russian summary at the end. Moreover, presumably for the benefit of foreign readers, the Russian papers carried a French or German summary.[18] However, the latter policy died away at around the same time as the use of foreign languages more generally: as a part of the 'patriotic campaigns' of the late 1940s, the Academy of Sciences decreed that Soviet journals would no longer translate abstracts or tables of contents into foreign languages.[19] Naturally, we see this decree taking effect in other periodicals that had adopted similar language policies to those of

---

[16]"Советская математика может и должна иметь журнал международного значения." (Anon 1931b).

[17]I record here only the numbers of papers in Russian and in languages other than Russian, but I hope elsewhere to analyse the distribution of the specific languages used. For the time being, suffice it to say that the foreign languages represented in Fig. 4.1 are, in various proportions, English, French, German, and Italian.

[18]Indeed, although it was considerably rarer than the converse, some Western journals also carried Russian summaries of papers: the *International Journal of Earth Sciences*, for example, began to provide these shortly after the launch of Sputnik (Montgomery 2013, p. 92).

[19]See, for example, Krementsov (2007, p. 61) or Gordin (2015, p. 225). English summaries were, however, reintroduced in certain contexts around a decade later: see Herner (1958).

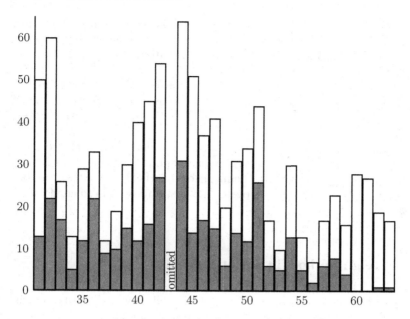

Fig. 4.1 Numbers of papers published in vols. 31–63 of *Matematicheskii sbornik* (in the numbering of the original series), showing both the total, and the number in foreign languages (shaded). Volume 43, whose fifth issue features papers from the First International Topological Conference, is omitted

Matematicheskii sbornik. Take, for example, the Kharkov-based journal *Soobshcheniya   Kharkovskogo   matematicheskogo   obshchestva   (Сообщения Харьковского математического общества = Communications of Kharkov Mathematical Society)*, which published papers in both French and German, alongside Russian and Ukrainian, with the explicit intention of boosting foreign readership.[20] In fact, almost the entirety of volumes VI–XIII (1933–1936) of the *Soobshcheniya* were published in German, French and (occasionally) English. Papers in French, German or English carried a summary in Russian or Ukrainian, whilst Russian and Ukrainian papers featured a French, German or (again, very occasionally) English summary. Here again, however, the policy of employing foreign languages faded away in the second half of the 1930s, as indeed did that of using Ukrainian: by the 1940s, almost all papers in the *Soobshcheniya* were in Russian.

The brief comments here concerning the *Soobshcheniya* raise the issue of the other languages of the Soviet Union. Although all member republics of the USSR used Russian to some degree, scientific publications did appear, at times, in other national languages: the above-mentioned use of Ukrainian in the *Soobshcheniya*, for example. Thus, when surveying the Soviet scientific literature, it is not unusual to encounter Ukrainian, Armenian, Estonian, and Georgian, amongst many other

---

[20]See Marchevskii (1956), and also the comments on this journal in Hollings (2014, p. 340).

languages. However, a Russian summary was typically appended to any paper in a national language—like Westerners, native Russian speakers typically had little facility with the USSR's other languages (Medvedev 1979, p. 157). Although intended as aids to other Soviet readers, such summaries also helped Western readers—even if Russian ability was not widespread in the West, it was nevertheless more readily available than expertise in some of the other languages of the USSR. Further comments on the other languages of the Soviet Union may be found in Sect. 4.4.

The near-exclusive use of Russian in Soviet scientific materials from the late 1940s onwards, labelled as "misplaced linguistic pride" by the more forthright (and 'English-promoting') of Western commentators (Garfield et al. 1986), thus became something for Western readers to overcome: their efforts to do so are dealt with in Sects. 4.4 and 4.5.

4.3 Foreign Authors in Soviet Journals

Throughout Chap. 2, I discussed the extent to which Soviet scientists were able to publish their work abroad, but I have so far said little about the opposite situation: Westerners publishing in Soviet journals. On the whole, this was somewhat rare, but not entirely unknown. As noted in Sect. 4.2, for example, some of the foreign-language papers published in *Matematicheskii sbornik* were in fact by foreign authors—Fig. 4.2 presents some data by way of illustration.[21] The numbers represented in Fig. 4.2 are certainly not huge, but they do affirm a foreign participation in *Matematicheskii sbornik*, a participation that the editors actively encouraged—in an editorial of 1931, we find the following bald statement:

> The editors invite the cooperation in the journal of foreign scholars sympathetic to the Soviet Union.[22]

Indeed, similar sentiments were expressed upon the launch of the journal *Uspekhi matematicheskikh nauk* (*Успехи математических наук* = *Progress of the Mathematical Sciences*) in 1936: not only that foreign authors would contribute to the journal, but, moreover, that they would provide a window onto foreign mathematical activities. In an editorial in the journal's first issue, for example, we find the following:

---

[21] See also the comments in Demidov (1996, pp. 140, 142). Much as in Fig. 4.1 (see note 17 on p. 80), I record here only the numbers of papers by foreign authors, but I hope elsewhere to analyse the distribution of the specific nationalities of authors (where by 'nationality' I mean country of stated affiliation, rather than country of origin). For the time being, suffice it to say that the nationalities represented in Fig. 4.2 are, in various proportions, American, British, Bulgarian, Czechoslovakian, Danish, Dutch, French, German, Hungarian, Italian, Lithuanian, Norwegian, Polish, Swiss, Turkish, and Yugoslavian.

[22] "Редакция приглашает сотрудничать в журнале иностранных ученых, сочувствующих Советскому союзу." (Anon 1931a).

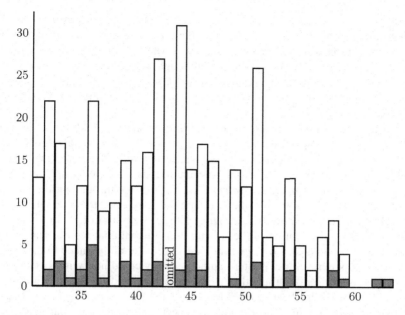

Fig. 4.2 Numbers of papers in foreign languages in vols. 31–63 of *Matematicheskii sbornik* (in the numbering of the original series), showing both the total number of foreign-language papers, and the number by foreign authors (shaded), based on stated affiliations. Volume 43, whose fifth issue features papers from the First International Topological Conference, is omitted

We have received a number of kind agreements of foreign mathematicians to give information on the mathematical work of some foreign mathematical centres: in this issue, for example, are printed informative articles by S. Lefschetz (USA) and A. Weil (France).[23]

Lefschetz contributed two articles (one in the above-mentioned issue, and another 2 years later) on mathematical activities at his home institution of Princeton (Lefschetz 1936, 1938), whilst Weil wrote on mathematics in France and in India, in which country he had spent some time (Weil 1936a, b). The articles of Lefschetz and Weil all appeared in Russian, and it seems likely that they were written in that language, rather than having been translated by someone else: Lefschetz certainly knew Russian (his parents were Russian: see Hodge 1973),[24] whilst Weil appears to have had at least a little Russian (Weil 1992, p. 109).

Lefschetz and Weil, however, were rare exceptions: the majority of Western authors who published in Soviet journals did so in languages other than Russian.

---

[23]"Далее следует отдел, посвященный информации о математической жизни. Нами получено любезное согласие ряда иностранных математиков давать информацию о математической работе некоторых иностранных математических центров: в настоящем выпуске, например, печатаются информационные статьи С. Лефшеца (США) и А. Вейля (Франция)." (Anon 1936, p. 4).

[24]See also note 36 on p. 19.

With regard to *Matematicheskii sbornik*, foreign interest appears to have increased, as the editors had hoped, as a result of the journal's language policy of the 1920s. Thus, as the journal's use of Western languages waned, so too did its foreign participation. It is interesting, however, that, although a shadow of its former extent, foreign participation did not drop away to zero, for there was the occasional Western author who contributed a paper in Russian: the British mathematician F.V. Atkinson, for instance.[25]

So far as I have been able to determine, there were no official bars to Western authors publishing in Soviet journals. One might suppose that Soviet editors would have been nervous of taking such submissions, for fear of official criticism, but the evidence does not seem to support this speculation: as we have seen, foreign contributions continued to appear in *Matematicheskii sbornik* throughout the years of Stalin's purges in the 1930s. On the contrary, we may offer the alternative speculation that the Soviet authorities recognised the submission of Western work to Soviet journals as a validation of the sought-after international repute. The decline in foreign contributions to Soviet journals was almost certainly linked to those journals' language policies: most Western authors simply could not write Russian well enough. Dissatisfaction with lengthy review processes, and unilateral decisions of Soviet editors, have also been cited.[26] Moreover, the near-exclusive use of Russian in Soviet journals may also have made them seem much more insular to Western authors: work published there simply would not reach a wide international audience.

4.4   Russian-Language Ability Amongst Western Scientists

As has been well-documented (see, for example, Montgomery 2013), during the 20th century, the greater part of (Western) scientific literature was published in German, French and English, with slight variations from discipline to discipline. Thus, the scientists of the 20th century (certainly those of Western Europe and North America), would usually have been able to read the publications that were relevant to them, provided they had some small grasp of German, French and English. A need to read material in other languages, such as Italian, might sometimes arise, but these remained the dominant three—hence the use of these languages in *Matematicheskii sbornik*. Moreover, the use of these languages would probably not have been to the detriment of Russian readers, since the principle that a 20th century scientist ought to have a knowledge of at least one of German, French and English seems to have held amongst scientists in the Soviet Union also. Western visitors to the USSR often

---

[25] See, for example, Atkinson (1951); see also Mingarelli (2005).

[26] See, for example, Schweitzer (1989, p. 181) or Medvedev (1979, p. 154). Interestingly, the delays noted by Medvedev for Soviet biological journals in the mid-1970s (1 year from submission to publication) do in fact compare quite favourably with the delays often encountered in modern academic publishing.

noted their hosts' facility with Western languages[27]: foreign languages (English, in particular) were widely taught in Russian schools.[28] At one point, the passing of a foreign language examination was made a requirement for would-be Soviet travellers (Medvedev 1979, p. 207).

On the other hand, as we have already observed, a knowledge of Russian, though by no means unknown, was not particularly widespread amongst Western scientists, for the entire span of the 20th century.[29] Although efforts were made now and then to increase Russian abilities amongst scientists, such activities often came up against the often-observed complacency of native English speakers, and their unwillingness to learn another language; in the United States, this issue was exacerbated by the fact that Americans were

> accustomed to crossing the continent in a jet plane ... and to finding the same language upon arrival. (Kertesz 1974, p. 86)

The false sense of security thus created meant, moreover, that few US scientists, travelling to foreign conferences,

> [found] it strange that they [could] present their papers in English while their Dutch or Norwegian hosts [did] not use their native tongues, even in their own country. (Kertesz 1974, p. 86)

Hence, the inaccessibility of materials in Russian was often acknowledged (see, for example, Bell 1933), but few scientists acted to remedy the situation. The shock engendered by the launch of Sputnik in 1957, however, besides driving a desire to learn more about Soviet scientific and technical developments (Sect. 3.5), also led to the very practical related wish to gain a greater understanding of the Russian language in order to be able to read the work of Soviet colleagues.[30] Russian-language courses were therefore arranged for scientists at many Western universities.[31] One of the results of this, as the author Masha Gessen has commented in the mathematical context, is the fact that

> there is a generation of American mathematicians who are more likely than not to possess a reading knowledge of mathematical Russian. (Gessen 2011, p. 7)

In many cases, however, the efforts on the part of scientists to learn Russian were somewhat half-hearted, and the courses that were taught within universities often left something to be desired—they were not necessarily taught by qualified language teachers, but simply by Russian expatriates. Thus, the broad understanding

---

[27] See, for example, Bockris (1958) or Berger (1963).

[28] See, for example, Chaitkin (1945), O'Dette (1957) or Gordin (2015, p. 222); see also the conclusions of the report Litchfield et al. (1958, Sect. 11).

[29] See, for example, the comments on this matter in Benedict (1909).

[30] See, for example, the comment in Lefschetz (1949); see also Large (1983, p. 45). On the broader impact of Sputnik on US education, see Douglass (2000).

[31] See Hanson (1962, Chap. 2). In the American context, see Gordin (2015, p. 228), where it is emphasised that the 'scientific Russian' learnt by US scientists was "a more docile, friendlier beast" than literary Russian.

of Russian in the West seems to have remained quite low. Indeed, the lack of general success in the education of American (and, more generally, Western) scientists in the basic principles of the Russian language led one author to accuse Americans of merely "pretending" to learn Russian (Beyer 1965, p. 46).

Aside from the problems of learning a new language, a lack of enthusiasm born of prejudice (such as that discussed briefly in Sect. 4.1) may also have played a role in the inability of Western scientists to learn Russian *en masse*. The effects of the difficulty simply of overcoming cultural barriers, and the "innate fear of the odd alphabet" (Kertesz 1974, p. 86), should also not be underestimated; recall, for example, the Loren Graham quotation on p. 3. Indeed, from as early as 1945, we have the comments of Jacob Chaitkin, one of many commentators who tried to encourage American scientists to learn some basic Russian:

> The only cloak of mystery that envelops Soviet science is that of the Russian language. Is this language a true barrier, or is it merely a psychological obstacle that we ourselves have conjured up? (Chaitkin 1945, p. 301)

Chaitkin went on to note that

> [t]he Russians do not permit language to form a similar obstacle in their study of our scientific work. English is taught widely in the schools of the U.S.S.R. and is treated as second in importance only to the native languages of that country. There is no psychological barrier in the Soviets' attitude toward the study of English. Perhaps we may put it this way: English is studied by many Russians because there it is considered easy; Russian is studied by few Americans because here it is considered difficult. (Chaitkin 1945, p. 301)

If a Western scientist did try to read a Russian article, then he or she would probably have found that the terminology used by Soviet authors was essentially the same as that of Western languages—since the end of the 17th century, Russian scientists had engaged in the wholesale importation of technical words from Western European languages,[32] and thus, like much of the rest of Europe, employed a technical vocabulary with a firm Greek and Latin basis, although efforts were occasionally made to replace foreign technical loan-words by Russian equivalents.[33] Soviet emphasis on the Russian language, or at least its use alongside other national languages, meant that Western readers had only one exotic language to try to get to grips with. Indeed, the prominence of Russian also aided in scientific communication *within* the USSR, as Medvedev observed:

> The proceedings of republican academies ... started to be published in the local languages ... It was not unusual for the research institute in Estonia to receive an official letter written in the Uzbek language from the Uzbekistan Soviet Republic, and it could take months before anyone could be found who was able to read it. In retaliation, the reply would be written in Estonian, and this would create the same situation in Uzbekistan. ... After a few years of frustration, the Russian language again was made the official language for internal communication.[34]

---

[32] See Gouzévitch and Gouzévitch (2009, p. 364); see also UNESCO (1957, Sect. 7.5.13). On Russian chemical terminology, see Gordin (2015, Chap. 3).

[33] See Terpigorev (1950) and Crisp (1989, pp. 34–35).

[34] Medvedev (1979, p. 128); see also the comments in Medvedev (1971, p. 160). On Soviet language policy, see Kirkwood (1989) or Lipset (1967).

Western scientific readers were helped further by the appearance of a vast[35] number of Russian language guides and dictionaries: general outlines aimed to give Western scientists a basic grounding in Russian grammar,[36] whilst subject-specific word-lists provided the necessary technical terms.[37] Introductory guides to scientific Russian also appeared in the much less daunting form of short articles,[38] as did materials to help scientists who already knew Russian to convey this skill to their colleagues.[39] Some disciplines were better served than others: as indicated in Sect. 4.1, Soviet work in particular areas was deemed by some Westerners to be unworthy of consideration. In other disciplines, however, a strong desire to learn of Soviet contributions resulted not only in the translation of much Russian work (see the next section), but also in the production of several subject-specific language guides: mathematics[40] and chemistry[41] were particularly well-served in this regard. Indeed, one writer claimed that "chemists constitute[d] the largest group of scientists studying Russian" (Tolpin 1949, p. 27); chemists appear to have been amongst the first scientists to recognise the need to read Russian materials,[42] and thus enjoyed the first English translations of Soviet journals, as we will see in the next section. Besides the straightforward language guides, would-be Russian readers were also assisted by a variety of other resources, such as manuals on the tricky business of transliteration,[43] and also on how to navigate the often Byzantine systems of Russian abbreviations.[44]

Although English-Russian dictionaries were by no means the only scientific language guides that were produced,[45] these were nevertheless particularly common,

---

[35]The number was probably 'vast' because such technical guides are very quickly out of date; see, for example, the comments in Large (1983, p. 66).

[36]See, for example, Bray (1945), Perry (1950), Condoyannis (1959), Turkevich and Turkevich (1959), Ward (1960), Anon (1963a), Pertzoff (1964), Holt (1964), Warne (1964), and Alford and Alford (1970). For a manual for would-be translators of Russian scientific literature, see Zimmerman (1967).

[37]See, for example, Anon (1957), Konarski (1962), Emin (1963), Lambert (1963), Gitcigrat et al. (1963), and Kotz (1964, 1966). The compilation of a list of Russian technical terms was one of the goals of an Anglo-Soviet Scientific Collaboration Committee (Anon 1942), but it is not clear whether this was ever completed.

[38]See, for example, Melnechuk (1963), Anon (1963c) or UNESCO (1957, Sect. 2.8); see also the further references in note 41 below, which, although aimed at chemists, are useful for scientists more generally.

[39]See, for example, Tolpin (1945, 1949, 1964) or Frank (1947).

[40]See, for example, Anon (1950), Lohwater (1961), Nidditch (1962), Burlak and Brooke (1963), and Gould (1972). For a manual for would-be translators of Russian mathematical literature, see Gould (1966).

[41]See, for example, Callaham (1947), Hoseh and Hoseh (1964), Reid (1970), Perry (1944), Wiggins (1972), and Kiefer (1970).

[42]See, for example, Tolpin (1946), Soule (1955), Dostert (1955), and Wood (1966).

[43]See, for example, Shaw (1949) and Anon (1953); see also Large (1983, p. 72).

[44]See, for example, Rosenberg (1952) and Scheitz (1961).

[45]Other examples are Klinkovstejn and Znamenskij (1963), Macintyre and Witte (1956), Czerni and Skrzyńska (1962), Gould and Obreanu (1967), and Nihon Sūgakkai (1968).

probably because of the difficulties outlined in this section. Nor was the production
of such language guides a purely Western concern: similar such resources were also
published in the USSR.[46] In fact, the number of available scientific dictionaries, covering a range of languages and disciplines, and published in a variety of countries
grew so vast that it became necessary for certain official bodies to publish guides
listing them.[47]

As already noted, however, the Russian-language-learning materials surveyed
above had only a limited impact on Western scientists. We may speculate also that
a disinclination to go to the effort of deploying language skills (as touched upon in
Sect. 4.1) afflicted even those scientists who had managed to learn some Russian.
Thus, even as late as 1987, we find comments such as the following:

> there is reason to believe that the paucity of Russian-speaking scientists blunts the West's
> capacity to determine trends and developments in Soviet science. (Holden 1987, p. 113)

In addition, it seems that many Russian scientists were still only known in the West
through those of their works that appeared in Western languages (principally English) (Garfield 1990): both those written originally in such languages, and also those
that had been translated. The fact is that the Western scientists of the Cold War
found themselves spoilt by the availability of (mainly English) translations of Soviet
works, a subject that we turn to in the following section.

4.5   Translation of Scientific Works

As noted in Sect. 4.1, and also in earlier sections, a great deal of debate surrounded
the issue of producing translations of Soviet scientific works. As with the provision
of Russian language guides for scientists, the printed material on this matter is quite
extensive, ranging from the very general to the discipline-specific,[48] and deals with
many of the issues that we have touched upon here—most particularly, the question
of whether the Soviet literature on a given topic contained anything worth reading.
Although many of the texts on translation cited here take a very broad approach,
the translation of publications specifically from Russian remains a prominent thread
throughout. The handling of materials in other languages which, from the Western
perspective, are particularly alien (such as Chinese and Japanese) is also given a
great deal of attention. For the greater part of this section, I will focus upon the
translation of texts *into English*.

As with the general literature on the foreign-language barrier, commentaries on
the need for scientific translations say broadly the same things over and over again:

---

[46]See, for example, Shtokalo (1960) and Tonian (1965); see also DuS (1956).

[47]See, for example, Wiggins (1972), UNESCO (1957, Appendix 3), Marton (1964), and
Holmstrom (1951).

[48]See, for example, Casagrande (1954), Citroen (1959), Gingold (1964), Tybulewicz (1970), Scott
(1971), Gould and Stern (1971), Anderson (1978), Finlay (1979), and Miner (1980).

that efforts to teach Russian to Western scientists have proved largely ineffective, and so, given the USSR's *penchant* for publishing exclusively in Russian, the only avenue open to Western researchers, if they were to learn of Soviet advances, was to fund the translation of Soviet scientific works. The motivations of both fear and curiosity that we saw in Sect. 3.5 once again played a major role here, and were often backed-up by specific examples of the cost of failing to note Soviet developments: the loss of prestige with regard to Sputnik, for instance, or the estimated financial loss of $250,000 in connection with an initially unnoticed Soviet advance in information theory (O'Dette 1957, p. 580). Moving beyond the Soviet context for a moment, we note also a very well-publicised instance of mushroom poisoning in the United States in 1970, where deaths might have been prevented, had US doctors been aware of an antidote that had been proposed in the Czech medical literature some years earlier.[49] The need for translations of scientific works thus appeared to be clear, and it simply remained to settle the details: who was to produce the necessary translations, and what form should the translations take.[50] As we have already noted in Sect. 4.1, there were, at least as far as scientific journals were concerned, two options: the publication of translations of selected papers ('*ad hoc* translations'), or the cover-to-cover translation of entire journal issues. Both options were taken, in different contexts, but neither was without its problems.

The translation of Russian scientific materials into Western languages had in fact been underway since at least the beginning of the 20th century, although these were typically *ad hoc* in nature. Take, for example, the translations of Russian work on metabolism that were overseen by Francis Gano Benedict in the early years of the 20th century. These translations, however, were not circulated widely, but were lodged merely in the library of Benedict's laboratory—although their existence was advertised in the pages of *Science* (Benedict 1909). We have seen other examples of *ad hoc* translations in earlier sections: the wartime translations of medical materials, for instance (Sect. 2.4).

The *systematic* translation of certain Russian materials appears to have begun in the 1930s, when the Amkniga Corporation, a New York-based publisher, began to supply American readers with English translations of Russian books (Furaev 1974, English trans., p. 67). However, these were typically literary, rather than technical, works. Nevertheless, the early recognition that there was a market for translations of Russian technical materials seems to have been due to private individuals and firms, rather than to academic bodies. Thus, for example, in the mid-1940s, the American entrepreneur Earl Coleman tested the waters by first publishing translations of individual Russian scientific articles, before embarking upon the cover-to-cover translation of the Soviet *Journal of General Chemistry* (Журнал общей химии), which appeared in English from 1949 to 1993 as the *Journal of General Chem-*

---

[49] See Shephard (1973), Chan (1977), or Large (1983, p. 3). Indeed, a similar, but more recent, example is provided by the lack of Western knowledge of important Chinese papers concerning bird flu: see Montgomery (2013, p. 109).

[50] Another matter that received a great deal of attention was the question of the cost and efficacy of machine translations; for an early survey of progress in this direction, see Locke (1956). See also Large (1983, Chap. 6) and Gordin (2015, Chap. 8).

istry of the USSR, and continues today as the *Russian Journal of General Chemistry* (Coleman 1994; Gordin 2015, p. 252ff). This was the world's first cover-to-cover translation of an academic journal; during the 1950s, Coleman's Consultants Bureau added many more translated titles to their catalogue. Indeed, having seen that such translations were, on the whole, financially viable, other bodies (in particular, academic bodies) began to produce their own. In the United States, a great deal of funding for translations was provided by the National Science Foundation (NSF)[51]; by 1958, the NSF was involved in the cover-to-cover translation of 53 Soviet scientific journals (Anon 1958). Moreover, as part of the post-war blurring of the US military/civilian scientific divide (Wolfe 2013, Chap. 2), the Office for Naval Research similarly funded many scientific translation programmes, such as the 'American Mathematical Society Translations' series (Anon 1960b). The UK's Department for Education and Science was also a major funder of journal translation programmes, although, according to one figure of 1968, 85 % of translations of Soviet scientific journals originated in the United States (from a range of bodies) (Rangra 1968, pp. 8–9).[52] In a discipline-specific example, the same source observed that, at that time, one in five of the papers handled by *The Physics Review* was from a journal translated from Russian (Rangra 1968, p. 7). The scale of the proliferation of translations of Soviet journals during the 1950s and later can be seen in the numerous advertisements in the general scientific literature,[53] and in the lengthy lists of translated journals that appear in various sources.[54] Indeed, new translations continued to be launched right up to the end of the Soviet era.[55] Translations of Soviet scientific books were similarly produced by various organisations. As is perhaps appropriate, given the origins of the widespread translation of Russian scientific materials, the translation of books was handled by both academic and corporate entities: the New York-based publisher Chelsea, for instance, was responsible for many English translations of Russian scientific works. In the UK, Pergamon Press was involved, from 1958, in the translation of some of the publications of the Soviet Academy of Sciences (Korneyev and Timofeyev 1977, p. 57). In later decades, some translations from Russian also originated within the USSR, prepared by the All-Union Agency for Copyright (Всесоюзное агентство по авторским правам) (Kryuchkova 2001, pp. 411–412).

---

[51] See, for example, Anon (1956) and Adkinson (1967).

[52] For some details of translation activities in other countries, see Frank (1961).

[53] With regard to the early 1960s, for example, see Anon (1960a, 1963b) and Armstrong (1961). The January 1964 issue of *Science East to West* (no. 13, pp. 8–9), for instance, features advertisements of new cover-to-cover translations of Soviet journals on instrumentation for measurement, experimental techniques, and automatic control.

[54] See, for example, Hanson (1962, Chap. 3) or Rangra (1968, pp. 11–21). See also the graph in Gorokhoff (1962, p. 15) showing the increase in the number of cover-to-cover translations of Soviet journals.

[55] See, for example, Cantor (1983) (on the cover-to-cover translation of a Soviet materials science journal), Adams (1983) (geophysics), Colwell (1983) (remote sensing), Vickerman (1985) (adsorption), Mills (1985) (chemistry), Oliver (1987) (microbiology and biotechnology), and Robinson (1990) (oceanography).

It should be noted that Western translations of Soviet journals (and of scientific works more generally) appear, on the whole, to have been direct, and even included translations of the ideological articles[56] that had sparked concerns of postal censorship in the USA (see Sect. 3.2). The only real difference that I have observed between some translations and their originals has been in their titles, which have occasionally been modified (if only slightly) for reasons that are unclear.[57] In sticking strictly to the content of the originals, Western translations have also preserved some unpleasant features of the Russian originals, such as the anti-Semitic remarks directed by the Russian mathematician L.S. Pontryagin towards his American counterpart Nathan Jacobson (Pontryagin 1978).[58]

Western scientific works were also translated into Russian, though, it seems, largely on an *ad hoc* basis (Gordin 2015, p. 250): the better language skills of Soviet scientists may have rendered the systematic translation of Western sources less critical, although such translations had been produced from time to time since both before and after the October Revolution.[59] The level of censorship that cover-to-cover translations would have required may also have deterred the Soviet authorities from sanctioning any large-scale translation programmes. Thus, it appears that the only Western scientific publications that received Russian translations were longer (usually one-off) book-length works. Nevertheless, Soviet readers do appear to have received Russian versions of a comprehensive cross-section of the Western scientific literature (O'Dette 1957, pp. 579–580), although print-runs of these were typically limited and sold out very quickly (Gerovitch 2013, p. 183). As with the desire to participate in exchange programmes, the drive to translate Western sources probably stemmed not only from simple curiosity, but also from the desire to catch up with, and to surpass, the West in all areas. Thus, many prominent Western scientific texts were translated into Russian in the decades following the Second World War. For example, after Lysenko's downfall, there was an upsurge in the translation into Russian of non-Lysenkoist foreign genetics texts, in order to enable Soviet geneticists to catch up with advances elsewhere (Langer 1967).[60] A prominent producer of such translations was the Moscow-based publisher Izdatelstvo Inostrannoi Literatury (Издательство Иностранной Литературы = Publisher of Foreign Literature). As its name suggests, this was a publishing house devoted exclusively to the translation of foreign materials. The publisher Mir (Мир = World/Peace) served a similar purpose (Medvedev 1979, p. 63). It should be noted that Soviet

---

[56] Such as, for example, the articles listed in note 12 on p. 59.

[57] Compare, for example, the (literal English translation of) the original Russian title of Hewitt (1986) with the title of its English translation, as given in the bibliography of this chapter. Note, incidentally, that this is an article by a Western author writing in Russian in a Soviet journal.

[58] For the background to Pontryagin's remarks, see Lehto (1998, Sect. 10.1).

[59] For example, on the translation into Russian of Western texts on eugenics between 1900 and 1917, see Krementsov (2011, p. 65). On translation of genetics texts during the 1920s, see Todes and Krementsov (2010, p. 350).

[60] For details of some translations of major Western mathematical texts into Russian, see Hollings (2014, Table 2.6).

translators also paid attention to Chinese and Japanese works; indeed, many of these subsequently reached the West because of their treatment at Russian hands (Hamel 1964).

As with translations in the opposite direction, translations of Western scientific materials into Russian appear to have been direct. Even those resources whose contents were, from the Soviet point of view, ideologically dubious were nevertheless translated directly,[61] although an editorial foreword was often added to translations of such materials, in which the 'philosophical shortcomings' of the text were 'exposed' and condemned. I point, for example, to the 1949 Russian edition of Oswald Veblen and J.H.C. Whitehead's *The foundations of differential geometry*, published by Izdatelstvo Inostrannoi Literatury as *Основания дифференциальной* [sic] *геометрии*,[62] where the editors took exception to the abstract approach to geometry that the authors had propounded (Hollings 2014, Sect. 10.4). Other complaints that we find in such prefaces concern the underrepresentation of the work of Soviet authors in the original text (Hollings 2014, Sect. 12.1.3).

To return to Western translations of Soviet works, we have already noted the debate surrounding these. Feelings concerning cover-to-cover translations in particular appear to have been mixed, with major concerns over the cost and timeliness of translations: one estimate (with a physics bias) suggested that subscriptions to translated journals produced by academic bodies cost two to three times those of the originals (translations produced by commercial publishers cost even more—even as high as 28 times the price of the original), and that delays of between 6 and 12 months could be expected from the appearance in print of an issue of a Soviet journal to that of its English translation (Rangra 1968, p. 9). In certain contexts, however, it was nevertheless felt that, even if it meant that some translated papers would never be read, it was still more cost-effective to translate all the papers in a given journal than to select papers for translation on an individual basis (Tybulewicz 1970, pp. 56–57)—though this was by no means a universally-held view. Workers in some disciplines saw cover-to-cover translations as a blessing: they appear to have been well-regarded in mathematics, for example, where many of the translations of Russian journals that were launched during the Cold War continue to operate.[63] Many physicists appear to have been similarly in favour of complete translations of the relevant journals (Tybulewicz 1970, p. 55). Geophysicists, on the other hand, questioned the cost-effectiveness of such translations, and suggested that the publication merely of English abstracts of Soviet geophysics papers would be a better use of resources.[64] Western translations of certain Soviet biomedical journals were discontinued after concerns were raised about their usefulness—it was thought that it might be more worthwhile to translate parts of the German biomedical literature

---

[61] In the case of mathematics, for example, see Vucinich (2002, p. 22).

[62] Original text: Cambridge University Press, 1932.

[63] See, for example, the list of translated journals maintained by the American Mathematical Society: http://www.ams.org/msnhtml/trnjor.pdf (last accessed 26th May 2015).

[64] See Anon (1969), Groos (1970), and Garfield (1970); see also Hamblin (2000, p. 303).

into English instead (Bishop and Pukteris 1973). A few years later, a poll of native-English-speaking biomedical researchers asked whether they would like to see new English translations of Soviet journals: the response was a resounding 'no' (Abrams 1971). Other authors writing in a more general setting were also very critical of the cover-to-cover translation of Soviet journals, branding it a short-term solution to a long-term problem ("a temporary crutch has unfortunately become, as too many crutches do, a wasteful modus vivendi"), and denouncing the difficulties created for librarians—both bibliographic confusion, and the expense of needing to pur-chase both English translation and Russian original (Garfield 1972, p. 335). Indeed, amongst fervently English-language-oriented commentators, the widespread publi-cation of English translations of Soviet journals was criticised for having discour-aged Soviet scientists from learning to write in English, thus setting them apart from much of the rest of the world's scientific community (Garfield 1972, p. 335). Whatever its disadvantages, however, there can be no denying that the production of extensive translations of Soviet works has been a very successful way of conveying Soviet scientific literature to Western readers (more effective, certainly, than rely-ing upon those readers to learn Russian), and, indeed, has resulted in commercial success for the various corporate publishers who have engaged in the production of such journals.

It remains to address one final point concerning the translation of Russian-language materials: their visibility. It is not enough simply for there to exist a trans-lation of a given Soviet journal: its potential readers need to be told about it. We have already noted the presence of advertisements of new translated journals in such pub-lications as *Nature* and *Science East to West*. More generally, just as we did for the holdings of original Soviet materials in Western libraries (Sect. 3.5), we find in the literature various library guides to the availability of translations, ranging from the general to the specific.[65] Besides letting readers know about those resources that had already been published, such guides also informed the reader about the trans-lation services offered by such bodies as (in the UK) the British Library,[66] or the special libraries association Aslib.[67] Indeed, guides of these kinds appear to have been necessary in the extreme: just as we observed a general ignorance on the part of Western scientists as to the availability of Soviet resources, the bibliographical lit-erature similarly laments the general lack of knowledge of the translation and inter-library lending services that were available to help scientists (and academics more generally) access foreign materials that might be of interest to them (Wood 1967, p. 129). Bibliographical workers also recognised that, besides the many systemati-cally translated Soviet resources, there were also many one-off translations that had been produced, either by individual scientists with the necessary language skills, or through the above-mentioned translation services. They therefore began to make efforts to collate such translations by creating indices of unpublished translations,

---

[65] See, for example, Thompson (1955), Gorokhoff (1962), Anon (1959a), and Himmelsbach and Boyd (1968).

[66] See, for example, Wood (1974) and Chillag (1980).

[67] See, for example, Birch (1979) and Glover (1979).

where researchers might find out whether a paper that they were interested in had
already been translated (see, for example, Mackiewicz 1955). The need to set up
such 'translation pools' (such as, in the United States, the Library of Congress
National Translations Center: see Kertesz 1974, pp. 93–4) was another of the major
issues dealt with by the various sources on the foreign-language barrier that were
cited in Sect. 4.1.[68]

References

Abrams, F.: A market survey: translations of foreign language biomedical periodicals. RQ **10**(4),
    321–324 (1971)
Adams, R.D.: Mineral spread. Nature 305(5934), 6 Oct, 495 (1983)
Adkinson, B.W.: The role of translation in the dissemination of scientific information. In: Citroen,
    I.J. (ed.) Ten years of Translation, pp. 91–103. Pergamon Press, Oxford (1967)
Alford, M.H.T., Alford, V.L.: Russian-English Scientific and Technical Dictionary. Pergamon,
    Oxford (1970)
Ammon, U.: Ist Deutsch noch internationale Wissenschaftssprache? Englisch auch für die Lehre
    an den deutschsprachigen Hochschulen. Walter de Gruyter (1998)
Ammon, U.: Language planning for international scientific communication: an overview of ques-
    tions and potential solutions. Current Issues in Language Planning 7(1), 1–30 (2006)
Anderson, J.D.: *Ad hoc* and selective translations of scientific and technical journal articles: their
    characteristics and possible predictability. J. Amer. Soc. Inform. Sci. **29**(3), 130–135 (1978)
Anon: London Maritime Conference, 1908–1909. Ministry of Foreign Affairs, Saint Petersburg
    (1910) (in Russian)
Anon: From the editor. Mat. sb. 38(1–2), 0 (1931a) (in Russian)
Anon: Soviet mathematicians, support your journal! Mat. sb. 38(3–4), 1 (1931b) (in Russian)
Anon: From the editors. Uspekhi mat. nauk, no. 1, 3–4 (1936) (in Russian)
Anon: Anglo-Soviet Scientific Collaboration Committee. Nature 150(3801), 5 Sept, 285–286
    (1942)
Anon: Russian-English Vocabulary with a Grammatical Sketch. Office of Naval Research, Wash-
    ington, DC/Amer. Math. Soc. (1950)
Anon: The Transliteration of Russian, Serbian and Bulgarian for Bibliographical Purposes. Royal
    Society, London (1953)
Anon: National Science Foundation Russian Translation Programme. Nature 178(4524), 14 Jul, 70
    (1956)
Anon: Russian-English Glossary of Nuclear Physics and Engineering. Consultants Bureau Enter-
    prises, New York (1957)
Anon: Russian scientific journals available in English. Nature 182(4636), 6 Sept, 632 (1958)
Anon: The Translations Bulletin. Nature 183(4661), 28 Feb, 581 (1959a)
Anon: Russian chemical journals. Nature 185(4717), 26 Mar, 891 (1960a)
Anon: Foreign science information: a report on translation activity in the United States. Notices
    Amer. Math. Soc. 7(1), 38–46 (1960b)
Anon: Problems of communicating scientific ideas. Nature 196(4859), 15 Dec, 1044–1045 (1962)
Anon: Soviet Russian. Scientific and Technical Terms. A Selective List. Library of Congress,
    Washington, DC (1963a)
Anon: Russian geology in translation. Nature 197(4865), 26 Jan, 340 (1963b)
Anon: Decoding Russian titles and sentences. Russian Tech. Lit., no. 12, Oct, 25–27 (1963c)

---

[68] See, for example, Chan (1976, p. 325) and UNESCO (1957, Sect. 4.6).

Anon: Geophysics: Russian translations. Nature 224(5221), 22 Nov, 750 (1969)

Anon: Sixty years of Soviet science. Nature 270(5632), 3 Nov, 1 (1977)

Armstrong, T.: Soviet polar research. Nature 191(4791), 26 Aug, 841–842 (1961)

Atkinson, F.V.: The normal solubility of linear equations in normed spaces. Mat. sb. 28(1), 3–14 (1951) (in Russian)

Barr, K.P.: Estimates of the number of currently available scientific journals. J. Documentation 23, 110–116 (1967)

Bell, E.T.: A suggestion regarding foreign languages in mathematics. Amer. Math. Monthly 40(5), 287 (1933)

Benedict, F.G.: Russian research in metabolism. Science 29(740), 5 Mar, 394–395 (1909)

Berger, K.: In the Soviet Union. ACLS Newsl. 14(2), 1–7 (1963)

Beyer, R.T.: Hurdling the language barrier. Physics Today 18(1), 46–52 (1965)

Birch, B.J.: Tracking down translations. Aslib Proc. 31(11), 500–511 (1979)

Bishop, D., Pukteris, S.: English translations of biomedical journal literature: availability and control. Bull. Med. Lib. Assoc. 61(1), 24–28 (1973)

Bockris, J.O'M.: A scientist's impressions of Russian research. The Reporter 18(14), 20 Feb, 15–17 (1958)

Bourne, C.P.: The world's technical journal literature: an estimate of volume, origin, language, field, indexing, and abstracting. Amer. Documentation 13(2), 159–168 (1962)

Bray, A.: Russian-English & English-Russian Scientific-Technical Dictionary, 2 vols. International Universities Press, New York (1945)

Burlak, J., Brooke, K.: Russian-English Mathematical Vocabulary. Oliver and Boyd, Edinburgh (1963)

Callaham, L.I.: Russian-English Technical and Chemical Dictionary. John Wiley & Sons, New York and London (1947); 2nd ed.: Russian-English Chemical and Polytechnical Dictionary (1962); 3rd ed. (1975); 4th ed.: Callaham, L.I., Newman, P.E., Callaham, J.R.: Callaham's Russian-English Dictionary of Science and Technology (1996)

Cantor, B.: Basic matters. Nature 305(5934), 6 Oct, 493 (1983)

Casagrande, J.B.: The ends of translation. Internat. J. Amer. Linguistics 20(4), 335–340 (1954)

Castro, R.: El español, lengua internacional. Yelmo 22, 5–10 (1975)

Chaitkin, J.: The challenge of scientific Russian. Sci. Monthly 60(4), 301–306 (1945)

Chan, G.K.L.: The foreign language barrier in science and technology. Intern. Lib. Rev. 8, 317–325 (1976)

Chan, G.K.L.: Mushroom poisoning, thioctic acid and the foreign language barrier. Aslib Proc. 29(6), 237–240 (1977)

Chillag, J.P.: Translations and translating services at the Lending Division. Interlending Rev. 8, 136–137 (1980)

Citroen, I.J.: The translation of texts dealing with applied science. Babel 5, 30–33 (1959)

Coleman, E.M.: The mass production of translation — for a limited market. Publ. Research Quarterly 10(4), 22–29 (1994)

Colwell, R.N.: Distant detection. Nature 305(5934), 6 Oct, 496 (1983)

Condoyannis, G.E.: Scientific Russian: A Concise Description of the Structural Elements of Scientific and Technical Russian. Wiley, New York (1959)

Congrès: XI International Navigation Congress, Saint Petersburg (1908). Proceedings of the congress. Lectures and communications by foreign members of the congress on issues relating to maritime navigation. Saint Petersburg (1910) (in Russian)

Couturat, L., Jesperson, O., Lorentz, R., Ostwald, W., Pfaundler, L.: International Language: Considerations on the Introduction of an International Language into Science. Translated by Donnan, F.G. Constable & Company, London (1910)

Crisp, S.: Soviet language planning 1917–53. In: Kirkwood (1989), pp. 23–45

Czerni, S., Skrzyńska, M. (eds.): Polish-English Technological Dictionary. Pergamon Press, Oxford/Wydawnistwa Naukowo-Techniczne, Warszawa (1962)

Demidov, S.S.: 'Matematicheskii sbornik' in 1866–1935. Istor.-mat. issled., no. 1(36), 127–145 (1996) (in Russian)

Dostert, L.E.: Foreign-language reading skill. J. Chem. Educ. **32**(3), 128–132 (1955)

Douglass, J.A.: A certain future: Sputnik, American higher education, and the survival of a nation. In: Launius, R.D., Logsdon, J.M., Smith, R.W. (eds.) Reconsidering Sputnik: Forty Years Since the Soviet Satellite, pp. 327–362. Harwood Academic, Amsterdam (2000)

DuS., G.: Scientific information in the U.S.S.R. Science 124(3223), 5 Oct, 609 (1956)

Emin, I.: Russian-English Physics Dictionary. Wiley, London/New York (1963)

Finlay, I.F.: Managing to meet translation needs. Aslib Proc. 31(11), 495–499 (1979)

Frank, A.: Translations of Russian scientific and technical literature in Western Countries. Rev. Documentation **28**(2), 47–55 (1961)

Frank, J.G.: Hints for teaching Russian. Modern Lang. J. **31**(1), 21–24 (1947)

Furaev, V.K.: Soviet-American scientific and cultural relations (1924–1933). Voprosy istorii, no. 3, 41–57 (1974) (in Russian); English trans.: Soviet Stud. Hist. 14(3), 46–75 (1975–1976)

Garfield, E.: Concerning cover-to-cover translation journals. Current Contents, no. 17, 29 Apr (1970); also appears. In: Essays of an Information Scientist, vol. 1. ISI Press, Philadelphia, PA. (1962–73)

Garfield, E.: Cover-to-cover translation of Soviet journals – a wrong 'solution' of the wrong problem. Current Contents, no. 19, 19 Jul (1972); also appears. In: Essays of an Information Scientist, vol. 1, pp. 334–335. ISI Press, Philadelphia, PA (1962–73)

Garfield, E.: Talking science. Nature 303(5917), 9 Jun, 554 (1983)

Garfield, E.: The Russians are coming! Part 2. The top 50 Soviet papers most cited in the 1973–1988 Science Citation Index and a look at 1988 research fronts. Current Contents, no. 25, 18 Jun, 3–13 (1990); also appears. In: Essays of an Information Scientist, vol. 13, pp. 216–226. ISI Press, Philadelphia, PA (1990)

Garfield, E., Small, H., Vladutz, G.: At the information exchange. Book review: 'Scientific Communications and Informatics' by A. I. Mikhailov, A. I. Chernyi and R. S. Giliarevskii. Translated by Robert H. Burger. Information Resources Press, Arlington, VA, 1984. Nature 319(6051), 23 Jan, 272 (1986)

Garfield, E., Welljams-Dorof, A.: Language use in international research: a citation analysis. Ann. Amer. Acad. Political Social Sci. **511**, 10–24 (1990)

Gerovitch, S.: Parallel worlds: formal structures and informal mechanisms of postwar Soviet mathematics. Hist. Sci. **22**(3), 181–200 (2013)

Gessen, M.: Perfect Rigour: A Genius and the Mathematical Breakthrough of the Century. Icon Books (2011)

Gingold, K.: Translations for the U.S. scientist. Chem. Eng. News, 17 Aug, 88–96 (1964)

Gitcigrat, E.E., Pinkevich, A.A., Utkina, I.A., Vinogradova, L.V.: English-Russian Dictionary on Exploration Drilling. Gostoptekhizdat, Leningrad (1963)

Glover, W.: Services provided by Aslib relating to translations. Aslib Proc. **31**(11), 525–529 (1979)

Gordin, Michael D.: Scientific Babel: The Language of Science from the Fall of Latin to the Rise of English. Profile Books (2015)

Gordon, M., Santman, A.: Language barriers, literature usage and the role of reviews: an international and interdisciplinary study. J. Inform. Sci. 3(4), 185–189 (1981)

Gorokhoff, B.I.: Providing U.S. Scientists with Soviet Scientific Information. National Science Foundation Washington DC (1959); revised ed. (1962)

Gould, C.J., Stern, B.T.: Foreign technical literature: a problem of costs, coverage and comprehension. Aslib Proc. **23**(11), 571–576 (1971)

Gould, S.H.: A Manual for Translators of Mathematical Russian. Amer. Math. Soc. (1966); revised ed. (1991)

Gould, S.H.: Russian for the Mathematician. Springer (1972)

Gould, S.H., Obreanu, P.E.: Romanian-English Dictionary and Grammar for the Mathematical Sciences. Amer. Math. Soc. (1967)

Gouzévitch, I., Gouzévitch, D.: Introducing mathematics, building an empire: Russia under Peter I. In: Robson, E., Stedall, J. (eds.) The Oxford Handbook of the History of Mathematics, pp. 353–373. Oxford Univ. Press (2009)

Groos, O.V.: Cover-to-cover translations. Nature 225(5231), 31 Jan, 482–483 (1970)

Hamblin, J.D.: Science in isolation: American marine geophysics research, 1950–1968. Physics in Perspective 2(3), 293–312 (2000)

Hamel, G.A.: Some figures on translation activities in the U.S.S.R. of Chinese and Japanese scientific literature. Science East to West, no. 17, Dec, 16–18 (1964)

Hanson, C.W.: The Foreign Language Barrier in Science and Technology. Aslib, London (1962)

Herner, S.: American use of Soviet medical research. Science 128(3314), 4 Jul, 9–15 (1958)

Hewitt, E.: What Pavel Sergeevich Aleksandrov means to me. Uspekhi mat. nauk 41(6), 205–208 (1986) (in Russian); English trans.: What Pavel Sergeevich Aleksandrov did for me. Russian Math. Surveys 41(6), 247–250 (1986)

Himmelsbach, C.J., Boyd, G.E.: A Guide to Scientific and Technical Journals in Translation. Special Libraries Association, New York (1968)

Hodge, W.: Solomon Lefschetz. 1884–1972. Biogr. Mem. Fellows Roy. Soc. 19, 432–453 (1973)

Holden, N.: International scientific communication — old problems and a new perspective. Linguist 26(3), 110–116 (1987)

Hollings, C.: Mathematics across the Iron Curtain: A History of the Algebraic Theory of Semigroups. Amer. Math. Soc. (2014)

Holmstrom, J.E.: Bibliography of Interlingual Scientific and Technical Dictionaries. UNESCO, Paris (1951); 2nd ed. (1952); 3rd ed. (1953); 4th ed. (1961); 5th ed. (1969)

Holmstrom, J.E.: The foreign language barrier. Aslib Proc. 14(12), 413–425 (1962)

Holt, A.: Scientific Russian: Grammar, Reading, and Specially Selected Scientific Translation Exercises and an Extensive Glossary. Wiley (1964)

Hoseh, M., Hoseh, M.L.: Russian-English Dictionary of Chemistry and Chemical Technology. Reinhold, New York/Chapman & Hall, London (1964)

Hunter, P.S.: The foreign language barrier — stumbling block or stepping stone. Inform. Scientist 4(2), 65–70 (1970)

Hutchins, W.J., Pargeter, L.J., Saunders, W.L.: The Language Barrier: A Study in Depth of the Place of Foreign Language Materials in the Research Activity of an Academic Community. Postgraduate School of Librarianship and Information Science, University of Sheffield (1971a)

Hutchins, W.J., Pargeter, L.J., Saunders, W.L.: University research and the language barrier. J. Librarianship Inform. Sci. 3(1), 1–25 (1971b)

Jaramillo, F.: El español, lengua internacional. Yelmo 25, 42–43 (1975)

Kertesz, F.: How to cope with the foreign-language problem: experience gained at a multidisciplinary laboratory. J. Amer. Soc. Inform. Sci. 25(2), 86–104 (1974)

Kiefer, P.A., Jr.: On translating chemical Russian. J. Chem. Documentation 10(2), 119–124 (1970)

Kirkwood, M. (ed.): Language Planning in the Soviet Union. Macmillan/School of Slavonic and East European Studies, University of London (1989)

Klinkovstejn, G.I., Znamenskij, A.N.: German-Russian Autotransport Dictionary. Avtotransizdat (1963)

Kojevnikov, A.: The Great War, the Russian Civil War, and the invention of big science. Science in Context 15(2), 239–275 (2002)

Konarski, M.M.: Russian-English Dictionary of Modern Terms in Aeronautics and Rocketry. Pergamon, Oxford/New York (1962)

Korneyev, S.G., Timofeyev, I.A.: U.S.S.R. Academy of Sciences: relations with research institutions, scientists and scholars of Britain (1917–1975). In: Korneyev, S.C. (ed.) USSR Academy of Sciences: Scientific Relations with Great Britain, pp. 8–69. Nauka, Moscow (1977)

Kotz, S.: Russian-English Dictionary of Statistical Terms and Expressions and Russian Reader in Statistics. Univ. North Carolina Press (1964)

Kotz, S.: Russian-English Dictionary and Reader in the Cybernetical Sciences. Academic Press, New York (1966)

Krementsov, N.: In the shadow of the bomb: U.S.-Soviet biomedical relations in the early Cold War, 1944–1948. J. Cold War Stud. 9(4), 41–67 (2007)

Krementsov, N.: From 'beastly philosophy' to medical genetics: eugenics in Russia and the Soviet Union. Ann. Sci. **68**(1), 61–92 (2011)

Kryuchkova, T.: English as a language of science in Russia. In: Ammon, U. (ed.) The Dominance of English as a Language of Science: Effects on Other Languages and Language Communities, pp. 405–423. Mouton de Gruyter, Berlin (2001)

Lambert, M.: A Short Russian-English Dictionary of Terminology used in the Soviet Rubber. Plastics and Tyre Industries, Maclaren, London (1963)

Langer, E.: Soviet genetics: first Russian visit since 1930's offers a glimpse. Science 157(3793), 8 Sept, 1153 (1967)

Lapo, A.V.: Vladimir I. Vernadsky (1863–1945), founder of the biosphere concept. Internat. Microbiol. 4, 47–49 (2001)

Large, A.: The Artificial Language Movement. André Deutsch, London (1985)

Large, J.A.: The Foreign-Language Barrier: Problems in Scientific Communications. André Deutsch, London (1983)

Lefschetz, S.: Mathematical activity in Princeton. Uspekhi mat. nauk, no. 1, 271–273 (1936) (in Russian)

Lefschetz, S.: Mathematical activity in Princeton in 1935–1937. Uspekhi mat. nauk, no. 5, 251–253 (1938) (in Russian)

Lefschetz, S.: Mathematics. Ann. Amer. Acad. Political Social Sci. **263**, 139–140 (1949)

Lehto, O.: Mathematics without Borders: A History of the International Mathematical Union. Springer (1998)

Lipset, H.: The status of national minority languages in Soviet education. Soviet Stud. **19**(2), 181–189 (1967)

Litchfield, E.H., Mettger, H.P., Gideonse, H.D., Glennan, T.K., Harnwell, G.P., Malott, D.W., Murphy, F.D., Scaife, A.M., Sparks, F.H., Wells, H.B.: Report on Higher Education in the Soviet Union. Univ. Pittsburgh Press (1958)

Locke, W.N.: Translation by machine. Sci. American **194**(1), 29–33 (1956)

Lohwater, A.J.: Russian-English Dictionary of the Mathematical Sciences. Amer. Math. Soc. (1961)

London, I.D.: A note on Soviet science. Russian Rev. **16**(1), 37–41 (1957)

Louttit, C.M.: The use of foreign languages by psychologists. Amer. J. Psychol. **68**(3), 484–486 (1955)

Louttit, C.M.: The use of foreign languages by psychologists, chemists, and physicists. Amer. J. Psychol. **70**(2), 314–316 (1957)

Lyusternik, L.A.: 'Matematicheskii sbornik'. Uspekhi mat. nauk 1(1), 242–247 (1946) (in Russian)

Macintyre, S., Witte, E.: German-English Mathematical Vocabulary. Oliver and Boyd, Edinburgh (1956); 2nd ed. (1966)

Mackay, A.L.: Sources of Russian scientific information. Aslib Proc. **6**(2), 101–110 (1954)

Mackiewicz, E.: Commonwealth Index to Unpublished Translations and Aslib Panel of Specialist Translators. Aslib Proc. **7**(2), 69–70 (1955)

Marchevskii, M.N.: History of the Mathematics Divisions in Kharkov University during the 150 years of its existence. Zap. Khark. mat. obshch. **24**, 7–29 (1956) (in Russian)

Marton, T.W.: Foreign-Language and English Dictionaries in the Physical Sciences and Engineering: A Selected Bibliography 1952–1963. US Dept. Commerce (1964)

Medvedev, Zh.A.: The Medvedev Papers: The Plight of Soviet Science Today. Macmillan, London (1971)

Medvedev, Zh.A.: Soviet Science. Oxford Univ. Press (1979)

Melnechuk, T.: Decoding Russian sentences. Internat. Sci. Tech., no. 19, Jul, 64–71 (1963)

Mills, I.M.: Across the divide. Nature 317(6035), 26 Sept, 308 (1985)

Miner, E.: On the desirability of publishing translations. Scholarly Publ. **11**, 291–299 (1980)

Mingarelli, A.B.: A glimpse into the life and times of F. V. Atkinson. Math. Nachr. 278(12–13), 1364–1387 (2005)

Montgomery, S.L.: Does Science Need a Global Language?. Univ. Chicago Press (2013)

Neswald, E.: Strategies of international community-building in early twentieth-century metabolism research: the foreign laboratory visits of Francis Gano Benedict. Hist. Stud. Nat. Sci. 43(1), 1–40 (2013)

Nidditch, P.H.: Russian Reader in Pure and Applied Mathematics. Oliver and Boyd (1962)

Nihon Sūgakkai: Japanese-English Mathematical Dictionary. Iwanami (1968) (in Japanese)

O'Dette, R.E.: Russian translation. Science 125(3248), 29 Mar, 579–585 (1957)

Oliver, S.: On the small side East and West. Nature 329(6137), 24 Sept, 372 (1987)

Perry, J.W.: Chemical Russian, self-taught: I. Suggestions for study methods. J. Chem. Educ. 21(8), 393–398 (1944); II. The vocabulary problem. Ibid. 23(1), 22–27 (1946); III. Inorganic chemical nomenclature. Ibid. 23(3), 116–122 (1946); IV. Organic chemical nomenclature. Ibid. 24(1), 28–45 (1947); V. Russian grammar. Ibid. 24(2), 79–93 (1947)

Perry, J.W.: Scientific Russian: A Textbook for Classes and Self-Study. Interscience, New York (1950)

Pertzoff, V.A.: Translation of Scientific Russian. Exposition, New York (1964)

Pontryagin, L.S.: A short autobiography of L. S. Pontryagin. Uspekhi mat. nauk 33(6), 7–21, (1978) (in Russian); English trans.: Russian Math. Surveys 33(6), 7–24 (1978)

Rangra, V.K.: A study of cover to cover English translations of Russian scientific and technical journals. Ann. Lib. Sci. Documentation 15(1), 7–23 (1968)

Rathmann, F.H.: Soviet scientific literature. Science 128(3325), 19 Sept (1958)

Reid, E.E.: Chemistry through the Language Barrier: How to Scan Chemical Articles in Foreign Languages with Emphasis on Russian and Japanese. Johns Hopkins Univ. Press (1970)

Robinson, I.S.: Making waves. Nature 347(6293), 11 Oct, 596–597 (1990)

Rosenberg, A.: Russian Abbreviations: A Selective List. Library of Congress, Washington DC (1952)

Scheitz, E.: Russische Abkürzungen und Kurzwörter. VEB Verlag Technik Berlin (1961)

Schweitzer, G.E.: Techno-diplomacy: US-Soviet Confrontations in Science and Technology. Plenum Press, New York and London (1989)

Scott, P.H.: Technical translations: meeting the need. Aslib Proc. 23(2), 89–99 (1971)

Shaw, E.P.: Transliteration: a game for the library sleuth. Bull. Med. Lib. Assoc. 37(2), 142–145 (1949)

Shephard, D.A.E.: Some effects of delay in publication of information in medical journals, and implications for the future. IEEE Trans. Professional Comm. PC-16(3), 143–147 (1973)

Shtokalo, I.Z. (ed.): Russian-Ukrainian Mathematical· Dictionary. Vidav. Akad. nauk Ukr. RSR, Kiev (1960) (in Russian/Ukrainian)

Sloan, S S., Alper, J. (eds.): Culture Matters: International Research Collaboration in a Changing World. Summary of a Workshop. The National Academies Press, Washington, DC (2014)

Soule, B.A.: Language ability. J. Chem. Educ. 32(3), 112–114 (1955)

Svetlikova, I.: The Moscow Pythagoreans: Mathematics, Mysticism, and Anti-Semitism in Russian Symbolism. Palgrave Macmillan, New York (2013)

Terpigorev, A.M.: Problems of scientific and technical terminology, Vestn. Akad. nauk SSSR, no. 8, 37–42 (1950) (in Russian); English trans.: Current Digest of the Soviet Press, I I(45), 36–37 (1950)

Thompson, A.: Translations in the Science Museum Library. Aslib Proc. 7(2), 68–69 (1955)

Todes, D., Krementsov, N.: Dialectical materialism and Soviet science in the 1920s and 1930s. In: Leatherbarrow, W.J., Offord, D. (eds.) A History of Russian Thought, pp. 340–367. Cambridge Univ. Press (2010)

Tolpin, J.G.: Teaching of scientific Russian. Amer. Slavic East Europ. Rev. 4(1/2), 158–164 (1945)

Tolpin, J.G.: Russian in the chemist's curriculum. J. Chem. Educ. 23(3), 123–126 (1946)

Tolpin, J.G.: The present status of teaching Russian for scientists. Modern Lang. J. 33(1), 27–30 (1949)

Tolpin, J.G.: Surveying Russian technical publications: a brief course. Science 146(3648), 27 Nov, 1143–1144 (1964)

Tolpin, J.G., Danaczko, J., Jr., Liewald, R.A., Mayerle, E.A., Mayerle, E.H., Olszanski, E.B., Sekera, V.C., Zimmer, C.: The scientific literature cited by Russian organic chemists. J. Chem. Educ. 28(5), 254–258 (1951)

Tonian, A.O.: Dictionary of Mathematical Terms in the English, Russian. Armenian. German and French Languages. Akad. nauk Armyansk, SSR, Erevan (1965) (in Russian)

Tschirgi, R.D.: Should scientists communicate — and if so, with whom? Bull. Med. Lib. Assoc. 61(1), 1–3 (1973)

Turkevich, J., Turkevich, L.B.: Russian for the Scientist. Van Nostrand, Princeton, NJ (1959)

Tybulewicz, A.: Cover-to-cover translation of Soviet scientific journals. Aslib Proc. 22(2), 55–62 (1970)

UNESCO: Scientific and Technical Translating, and Other Aspects of the Language Problem. Documentation and Terminology of Science Series. UNESCO, Paris (1957)

Vickerman, J.C.: Just on the surface. Nature 317(6035), 26 Sept, 307 (1985)

Vucinich, A.: Soviet mathematics and dialectics in the post-Stalin era: new horizons. Historia Math. 29, 13–39 (2002)

Walsh, W.B.: Some judgments on Soviet science. Russian Rev. 19(3), 277–285 (1960)

Ward, D.: Russian for Scientists. Univ. London Press (1960)

Warne, E.J.D.: A Russian Scientific Reader. Allen & Unwin, London (1964)

Weil, A.: The mathematical sciences in France. Uspekhi mat. nauk, no. 1, 267–270 (1936a) (in Russian)

Weil, A.: Mathematics in India. Uspekhi mat. nauk, no. 2, 286–288 (1936b) (in Russian)

Weil, A.: The Apprenticeship of a Mathematician. Springer (1992)

Wiggins, G.: English-language Sources for Reference Questions Related to Soviet Science (With an Emphasis on Chemistry). Univ. Illinois Grad. School of Library Science Occasional Papers, no. 102, Jun (1972)

Wolfe, A.J.: Competing with the Soviets: Science, Technology, and the State in Cold War America. Johns Hopkins Univ. Press (2013)

Wood, D.N.: Chemical literature and the foreign-language problem. Chemistry in Britain 2(8), 346–350 (1966)

Wood, D.N.: The foreign-language problem facing scientists and technologists in the United Kingdom — report of a recent survey. J. Documentation 23(2), 117–130 (1967)

Wood, D.N.: Access to information in foreign languages — an experiment. BLL Rev. 2, 12–14 (1974)

Zikeev, N.T. (ed.): Scientific and Technical Serial Publications of the Soviet Union, 1945–1960. US Government Printing Office (1963)

Zimmerman, M.: Russian-English Translators [sic] Dictionary: A Guide to Scientific and Technical Usage. Plenum Press, New York (1967); 2nd ed. (1984); 3rd ed. (1992)

Chapter 5
Concluding Remarks and Points to be Pursued

Abstract We present here some concluding remarks, and points to be pursued. It is hoped that the detailed references provided throughout the book, in conjunction with the comments of the present chapter, will provide the impetus for further research in this area.

Keywords International scientific organisations · International scientific collaboration · The use of data in history of science

Although I have offered a very broad perspective on the issues of Cold War scientific communications, I nevertheless recognise that many of the topics touched upon here might be better dealt with separately, or in the context of particular disciplines. The existence of the various works cited in Chap. 1,[1] each dealing with East-West communications in specific branches of science, perhaps bears this out; the great value of subject-specific accounts lies in their ability to tell individual stories. My purpose, however, has been to point out the broad similarities experienced across disciplines.

Alongside the similarities, we have also seen some small differences between disciplines; these were much more in evidence in our discussions of language-related issues: we noted, for example, the glut of Russian-language guides and dictionaries aimed at mathematicians and chemists.[2] Indeed, one thing that emerges from Sects. 4.4 and 4.5 is that mathematicians and chemists were probably the most active accessors of the relevant Soviet scientific literature. The world-standing of Soviet mathematics provides an easy explanation of the first half of this proposition (Dalmedico 1997). However, as we have seen (Sect. 4.1), the Western view of Soviet chemistry was not always so favourable. Further research is thus required to explain Western chemists' apparent passion for the Soviet chemical literature.[3] Similarly, some detailed work might yet be done in connection with those disciplines whose Soviet versions Western scientists shunned, in order to determine whether the

---

[1] Specifically, on p. 1 and in the footnotes thereupon.

[2] See notes 40 and 41 on p. 87.

[3] As a further reference for the discussion of the Soviet chemical literature, see UNESCO (1957, Sect. 1.2.11). See also the comments in Gordin (2015, p. 217).

© The Author(s) 2016

C.D. Hollings, *Scientific Communication Across the Iron Curtain*,
SpringerBriefs in History of Science and Technology,
DOI 10.1007/978-3-319-25346-6_5

negative assessments of Soviet developments were justified, or simply the result of prejudice. Remaining in the realms of specific disciplines, there is enormous scope for further research into the functioning of international bodies and/or the organisation of international congresses. Within the present book, I have cited detailed studies of these aspects of the international research community for psychology (Rosenzweig 2000), physiology (Fenn et al. 1968), mathematics (Lehto 1998), crystallography (Kamminga 1989), archaeology (Babes and Kaeser 2009), and astronomy (Blaauw 1994)[4]—any other disciplines might be given similar treatments (indeed, as a further reference, see, for example, (Tikhomirov 1984) on geology). To pick up on a point raised in Sect. 2.6 in connection with astronomy, it remains to be explained why the USSR was so actively involved in the proceedings of the IAU at a time when it largely shunned other international scientific organisations.

Further questions arise for many of the minor points that I have but barely touched upon. For example, I mentioned the Anglo-Soviet Medical Committee in Sect. 2.4, but said only a little about its activities. In fact, little academic research appears to have been done on this Committee, in contrast to the small amount of literature on its American counterpart. It would be natural, for instance, to investigate just how effective the Committee was. Similar questions might also be asked about the SCR's Science Section (Sect. 2.2), and also about such bodies as the British Committee for Aiding Men of Letters and Science in Russia (Sect. 3.1).[5] In all this, there is a keen need for further Russian-language sources.

The provision of statistical evidence is another area in which further research might be conducted. At least some of what is generally know about Soviet science, or about levels of scientific communication, throughout this period is anecdotal, and is handed down as 'received wisdom'. I believe that it is desirable to supplement the existing sources with hard numerical data. Using tables and figures, I have attempted to do this on a very small scale in the present book, but more might yet be done. For instance, further data would serve to place my observations on Soviet 'local publication' (Sect. 3.3) on a firmer basis. Moreover, there is scope for expanding upon the figures I have given for the levels of foreign publication of Soviet mathematicians (Figs. 2.1, 2.2 and 2.3). How do these figures compare with those for other disciplines? Anecdotal evidence suggests that researchers in other areas did indeed send a lot of their work abroad, but is there any comparison to be made across different subject-areas? Moreover, the figures given for both foreign languages and foreign authors in *Matematicheskii sbornik* (Figs. 4.1 and 4.2) might be compared with similar figures for other journals (mathematical and otherwise). In this connection, there arises the question of the motivations of non-Soviet authors in their sending work to Soviet journals (as noted, for example, in Sect. 4.3). It can thus be seen that the use of data will not only serve to back up the many claims that can be made about international scientific communication during the Cold War, but will also lead to interesting new questions.

---

[4]Such volumes stand alongside those of a much more general nature, such as Greenaway (1996).

[5]The Anglo-Soviet Scientific Collaboration Committee and the American-Soviet Science Society, which have appeared here only in passing (respectively, in note 37 on p. 87, and in note 20 on p. 12) would similarly bear further investigation.

References

Babes, M., Kaeser, M.-A. (eds.): Archaeologists Without Boundaries: Towards a History of International Archaeological Congresses (1866–2006). Archaeopress, Oxford (2009)

Blaauw, A.: History of the IAU: The Birth and First Half-Century of the International Astronomical Union. Springer (1994)

Dalmedico, A.D.: Mathematics in the twentieth century. In: Krige, J., Pestre, D. (eds.) Science in the Twentieth Century, pp. 651–667. Harwood Academic, Amsterdam (1997)

Fenn, W.O., Franklin, K.J., Zotterman, Y. (eds.): History of the International Congresses of Physiological Sciences 1889–1968. Amer. Physiol. Soc. (1968)

Gordin, M.D.: Scientific Babel: The Language of Science from the Fall of Latin to the Rise of English. Profile Books (2015)

Greenaway, F.: Science International: A History of the International Council of Scientific Unions. Cambridge Univ. Press (1996)

Kamminga, H.: The International Union of Crystallography: its formation and early development. Acta Crystallogr. A **45**, 581–601 (1989)

Lehto, O.: Mathematics without Borders: A History of the International Mathematical Union. Springer (1998)

Rosenzweig, M.R.: History of the International Union of Psychological Science (IUPsyS). Psychology Press, Hove (2000)

Tikhomirov, V.V.: On the history of the international geological organisations. Voprosy istor. estest. tekhn., no. 3, 77–88 (1984) (in Russian)

UNESCO: Scientific and Technical Translating, and Other Aspects of the Language Problem. UNESCO, Paris (1957)

Index

© The Author(s) 2016
C.D. Hollings, *Scientific Communication Across the Iron Curtain*,
SpringerBriefs in History of Science and Technology,
DOI 10.1007/978-3-319-25346-6

Printed in the United States
By Bookmasters